ENDANGERED ENVIRONMENTS!

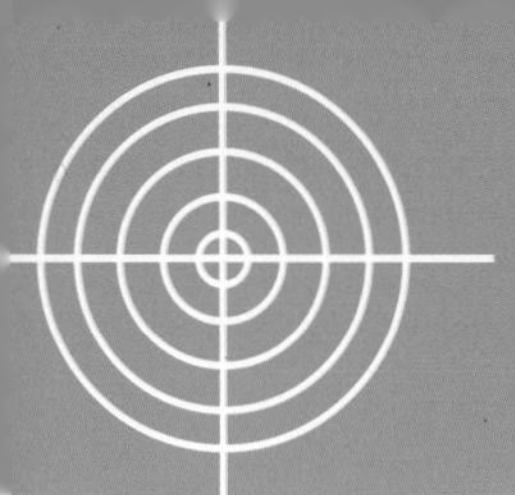

For a free color catalog describing Gareth Stevens' list of high-quality books, call 1-800-542-2595 (USA) or 1-800-461-9120 (Canada). Gareth Stevens' Fax: (414) 225-0377.

Library of Congress Cataloging-in-Publication Data available upon request from publisher. Fax: (414) 225-0377 for the attention of the Publishing Records Department.

ISBN 0-8368-1423-1

Exclusive publication in North America in 1996 by
Gareth Stevens Publishing
1555 North RiverCenter Drive, Suite 201
Milwaukee, Wisconsin 53212, USA

A LOVELL JOHNS PRODUCTION created, designed, and produced by Lovell Johns, Ltd., 10 Hanborough Business Park, Long Hanborough, Witney, Oxfordshire OX8 8LH, UK.

U.S. series editor: Patricia Lantier-Sampon

Printed in the United Kingdom

1 2 3 4 5 6 7 8 9 99 98 97 96

ENDANGERED ENVIRONMENTS!

Gareth Stevens Publishing
MILWAUKEE

CONTENTS

FOREWORD

Mark Collins, Director of the World Conservation Monitoring Centre.

In 1963, the IUCN Species Survival Commission, chaired by Sir Peter Scott, commissioned research and a series of books aimed at drawing to the attention of governments and the public the global threats to species. Sir Peter wanted more concerted action to address the problem of extinction. The first Red Data Book, published in 1969, was written by James Fisher, Noel Simon, and Jack Vincent. There had been earlier books that highlighted animals under threat and the possibility of extinction, the most important being written by G. M. Allen in 1942. The increasing threat to species and our knowledge of these threats has resulted in nearly 6,000 species being listed as threatened in the most recent IUCN Red List of Threatened Animals. (IUCN uses different categories of threatened species, of which the most crucial category is Endangered.)

Knowledge of the conservation status of species is required so priorities can be set and management actions taken to protect them. The original Red Data Books were global assessments of species. However, many of these globally threatened species are found in only one country, and it has become increasingly important for each country to assess its own species and decide which should be listed as Threatened. There are

Some of the many Red Data Books published since the first one appeared in 1969.

now various National Red Data Books covering substantial areas of the world.

The very fact that there are so many threatened species makes it very difficult to publish books on their status and distribution and, for some of them, we do not have detailed information. This *Endangered!* series aims to provide sound knowledge of 150 selected endangered animals and their natural habitats to a wider audience, particularly young people.

Our knowledge of threatened species can only be as good as the research work that has been carried out on them and, as the charts on this page show, the conservation status of much of the world's wildlife has not yet been assessed. Even for mammals, only about 55% of the species have been assessed. The only major group of which all species have been assessed are birds, and yet there are still large gaps in our knowledge of the status and trends in bird population numbers. However, their attractiveness and the interest shown in them by a great many people have improved the information available. Marine fish, despite their importance as a valuable food source throughout the world, tend to be assessed for conservation purposes only when their populations reach such a low point that it is no longer viable to catch them commercially. The 1994 IUCN Red List of Threatened Animals lists 177 endangered mammals representing 3.8% of the total number of mammal species and 188 birds representing 1.9% of the total number of species. Information on birds is compiled by BirdLife International.

The importance of identifying threatened species cannot be stressed enough. There have been many cases where conservation action has been taken as a result of the listing of species as endangered. The vicuna, a camel-like animal that lives in the high Andes of South America whose wool is said to be the finest in the world, was extremely abundant in ancient times but has been over-exploited since the European colonization of South America. By 1965, it was reduced to only 6,000 animals.

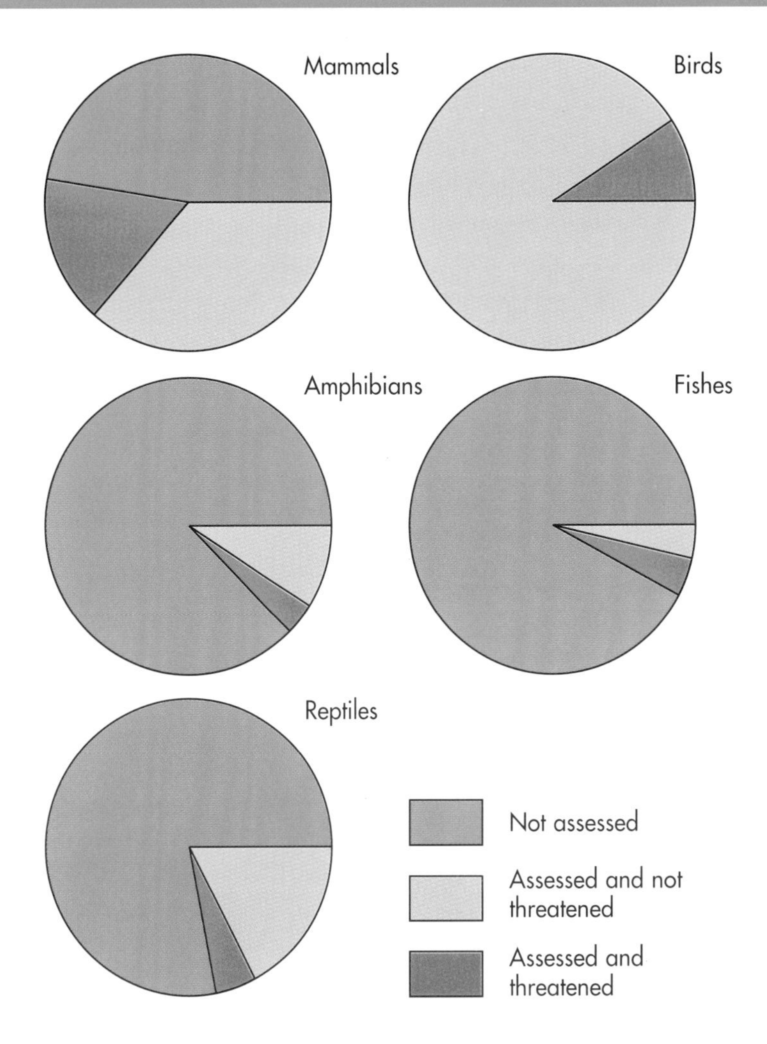

By protecting the vicuna from hunting and by establishing reserves, the population has steadily increased and is now in the region of 160,000. The vicuna is no longer an endangered species but is listed as Vulnerable. One high-profile endangered animal is the Indian Tiger, whose numbers had dropped to fewer than 2,000 in India when the first census was taken in 1972. Urgent conservation measures were taken, reserves were set up, and a great deal of expertise in their management has resulted in a population increase to its current level of about 3,250. Other measures included the halting of trade in tiger skins and other products such as bones and blood used in eastern traditional medicines. However, the other subspecies of tigers have not had this same protection, and their numbers are dwindling day by

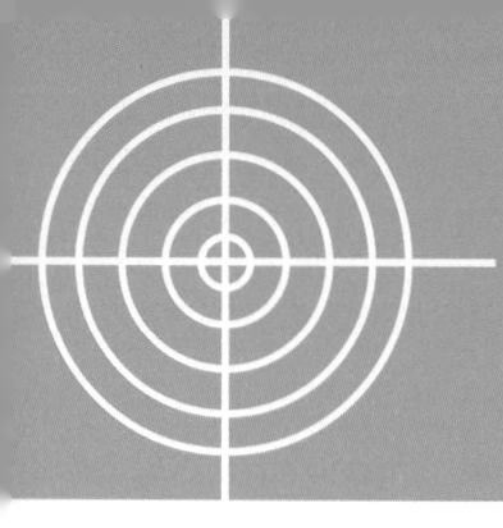

day. Gray whales were also endangered. They migrate down the west coast of North America from arctic waters to the coast of Mexico and southern California to mate, returning for the rest of the year to feed and give birth to their young. As their migration route was so well known, hunting was easy, and, as a result, their numbers had dropped to only a few hundred. Since hunting control measures began, the number of gray whales is now in excess of 21,000, and they are no longer listed as endangered.

WCMC

The World Conservation Monitoring Centre in Cambridge, in the United Kingdom, has been the focal point of the management and integration of information on endangered plant and animal species for more than fifteen years. WCMC's databases also cover the trade in wildlife throughout the world, information on the importance and number of areas set up to protect the world's wildlife, and a Biodiversity Map Library that holds mapped data on many of the world's important sites and ecosystems. It was IUCN, through its Species Survival Commission, that first established the World Conservation Monitoring Centre as its information database for species and ecosystems throughout the world. WCMC now carries on this role with the support of two other partners: the World Wide Fund For Nature and the United Nations Environment Programme.

IUCN — The World Conservation Union

Founded in 1948, The World Conservation Union brings together states, government agencies, and a diverse range of nongovernmental organizations in a unique world partnership: over 800 members in all, spread across some 125 countries. As a Union, IUCN seeks to influence, encourage, and assist societies throughout the world to conserve the integrity and diversity of nature and to ensure that any use of natural resources is equitable and ecologically sustainable. The World Conservation Union builds on the strengths of its members, networks, and partners to enhance their capacity and to support global alliances to safeguard natural resources at local, regional, and global levels.

Various organizations too numerous to mention help countries protect their wildlife. We urge you to support these organizations so the list of endangered species does not continue to grow. Your voice will be added to the many millions that are urging international cooperation for the protection and wise use of the wildlife that is such an important part of our natural heritage.

Headquarters of the World Conservation Monitoring Centre, Cambridge, England.

HABITATS OF THE WORLD

The surface of our planet Earth is divided into a number of major habitat types, called biomes. These are natural communities of plants and animals that cover a large area. There are ten main biomes on land and five in the sea, but these can be divided into many more. This book describes the twelve most important world habitats.

The biome that occurs at any one place depends mainly on its climate, especially the temperature and precipitation. The temperature difference between the heat of the equator and the cold of the poles causes winds that carry moisture from the oceans over the landmasses. Land heats up and cools down faster than the sea, so there is greater variation between summer and winter in the middle of continents than over the oceans. The combination of these effects, together with the type of soil and the topography, or shape, of the land, gives rise to the conditions that cause the different major habitats. In each, there is a natural community of plants and animals that are adapted for life in the habitat's physical conditions. For instance, species that live in polar habitats must cope with cold throughout the year, and those that live on the seashore must survive alternating wet and dry conditions with each tide.

Until a few hundred years ago, the world's habitats changed very slowly. Then two things happened that began to speed up the change. The human population began to increase rapidly, and technology started to make an impact on the world as people created new industries. The growing human population needed more crops to provide food and clothing. The new industries needed more raw materials, such as coal, iron, rubber, wood, and many minerals. Both humans and their industries caused environmental pollution. Even so, there were still huge areas of wilderness that seemed to escape destruction.

The rate of change has increased enormously in the last one hundred years. Human populations are growing faster than ever. One hundred and fifty years ago, the human population of the world was 1 billion. Fifty years ago, it was 2 billion. Twenty years ago, it was 4 billion. Now there are over 6 billion people. One hundred million more are born every year in South America. All these people need food, a place to live, and fuel for cooking and heating. In developed countries, people want automobiles, vacations, and many other luxuries. The average inhabitant of Canada uses 145 times as many of the world's resources as a person in Zaïre. So it is not surprising that the world's wild places are rapidly disappearing. People in developed countries want to preserve their quality lifestyle, and those in poor countries want to improve theirs. This can only be accomplished at the expense of the natural environment.

People all over the world worry about the world's disappearing habitats. They realize that conservation involves more than preserving endangered animals, such as giant pandas, whales, and butterflies. Species can often be preserved by banning hunting and keeping the animals in reserves. Preserving habitats is even more difficult. Most countries have only one percent or less of their land designated for reserves and national parks. Land is increasingly in demand for agriculture, industry, and human settlement. There is less hope for natural wild places as the human population continues to increase. Even many protected reserves and parks are endangered.

POLAR REGIONS

The difference between the Arctic and the Antarctic is that the Arctic is an ocean surrounded by the landmasses of America, Asia, and Europe, and the Antarctic is a continent surrounded by the Southern Ocean (also referred to as the Antarctic Ocean, or Southern Pacific, Southern Indian, or Southern Atlantic oceans). In the cold Antarctic and Arctic, summers are very short and winters are very long. Polar lands are also deserts because they get very little snow throughout the year. When the snow melts, the ground is mainly bare except for streams running from glaciers or snowbanks or where meltwater gathers and the ground is waterlogged.

The desolate beauty of the polar regions is illustrated by this picture of icebergs grounded in shallow waters between the open sea and the Riiser-Larsen Ice Shelf in Antarctica.

The Arctic habitat is called the tundra, and it lies north of the coniferous forests of the taiga (page 18). During the short summer, the ground surface thaws, and plants can grow. Over one hundred kinds of flowering plants grow on the tundra, and, for a short time each year, the ground is covered with colorful flowers. The plants are low-growing, so they are covered with snow at the beginning of winter. This protects them from damage by drifting snow. Arctic willow and arctic birch are "trees" that grow only a few inches (centimeters) tall. Insects appear briefly in the summer, along with clouds of mosquitoes, midges, butterflies, and bumblebees.

Birds migrate to the tundra for the summer to feed on the plants and insects. Most are waders, geese, and

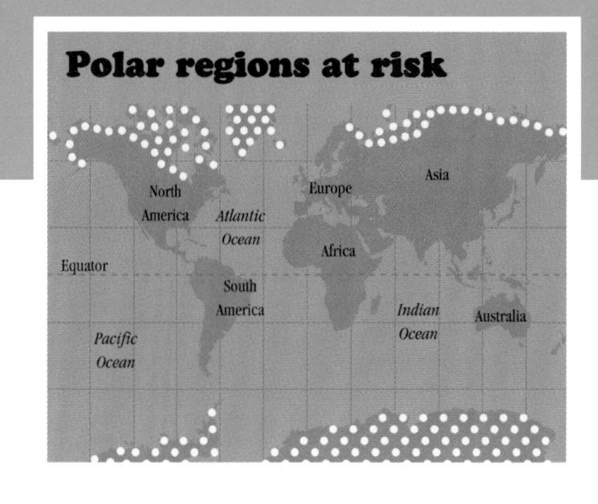

ducks. They arrive when the snow begins to melt on the tundra, and they rear their families quickly so they can fly south again before winter begins. Only ptarmigan that burrow into the snow, ravens that scavenge for food, and snowy owls that hunt small animals remain in the Arctic during winter. Plant-eating mammals include reindeer, caribou, muskoxen, Arctic hares, and lemmings. Lemmings are relatives of the voles and live in burrows underground or under the snow in winter. Their numbers vary enormously from year to year. When lemmings are abundant, the animals that hunt them — Arctic foxes, snowy owls, and long-tailed skuas — also increase in numbers. The largest predators are wolves and polar bears. The mighty polar bears spend most of their time on the frozen sea where they hunt seals.

The Arctic Ocean is frozen for most of the year, but the ice breaks into ice floes in summer. Seabirds like auks and guillemots nest on cliffs and dive for fish in open water. Ringed seals, bearded seals, and harp seals are fish-eaters, and they bear their pups on the ice. Walruses search for clams on the seabed. The Arctic Ocean is also home to bowhead and Greenland right whales, beluga or white whales, and narwhals.

The native Inuit and others have hunted in the Arctic for thousands of years. In northern Europe and Asia, the Sami (Lapps), Chukchi, and others lived by herding reindeer. People from outside the Arctic began to visit as explorers, traders, and hunters. They wanted the fur of polar bears, seals, and foxes; the ivory tusks of walruses; and the valuable items that came from whales. As a

From its starting point at Prudhoe Bay in Alaska, the Trans-Alaska pipeline stretches into the distance across a polar landscape. Oil drilling in the Arctic has been a major factor in the destruction of polar lands.

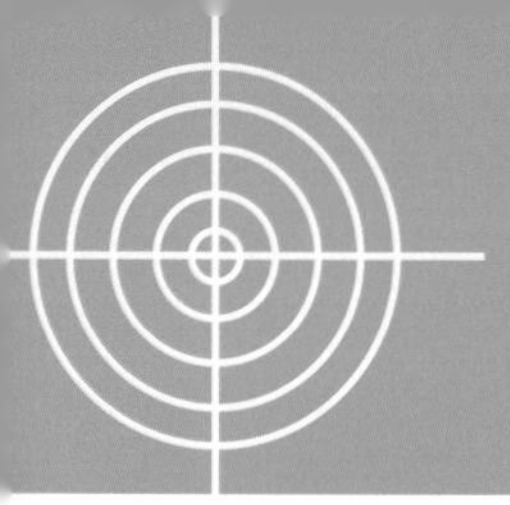

This beautiful blue iceberg in Antarctica provides a resting place for chinstrap penguins. Blue icebergs form from ancient, compressed ice and are extremely rare.

result, many animal populations became scarce. Nations with territories in the Arctic have finally united to save polar bears. The bears are fully protected in some places, although local people can hunt a certain number every year.

The Arctic is currently being exploited for oil and minerals. Scientists worry that these industries will permanently harm the arctic environment. Heavy vehicles make deep tracks in tundra soil unless they travel in winter, and the wildlife is disturbed. There are fears that oil pipelines will interfere with herds of migrating caribou. In 1977, the 790-mile (1,270-kilometer) Trans-Alaska pipeline started to carry oil. New roads and airports allow many more people into the Arctic as tourists, hunters, and pollution increase. However, some arctic nations have strict laws for controlling visitors and keeping the tundra clean. Oil spills from broken pipelines on the tundra and wrecked oil tankers in the sea present the worst threats. Cleaning up any environmental disaster in such remote places is very difficult, and damage may take centuries to repair.

Although both are cold places, the Antarctic is different from the Arctic. The continent of Antarctica is almost entirely covered with snow and ice, and it does not warm up in summer as much as the arctic tundra. Antarctica has snow-free places around the coastline and on islands in summer only. Whereas animals and plants spread into the Arctic tundra from lands farther south, Antarctica is separated from other continents by the huge expanse of Southern Ocean, and very few plants and animals manage to reach it. Only two species of flowering plants, together with some mosses and lichens, a few insects, and other invertebrate animals, flourish on the Antarctic continent.

On the other hand, the Southern Ocean that surrounds Antarctica is rich in life. Penguins and other seabirds, whales, and seals make their homes there. These animals feed on fish, squid, and crustaceans. The main food for most animals is krill, a type of shrimp that lives in vast swarms. During the summer, when the frozen sea thaws, minute floating plants called phytoplankton grow and multiply. The krill and other planktonic animals gather at the surface to feed on the phytoplankton. They, in turn, attract the fish, squid, penguins, seals, and other animals. Whales, like the blue, fin, minke, and humpback whales, migrate southward from their breeding places in warmer seas to spend the antarctic summer feeding on the swarms of plankton.

Antarctica is now the largest wilderness in the world. Explorers and scientists were the only visitors until about thirty years ago, when tourists began to arrive. Scientists believe that even a few people can damage the antarctic environment. At the present time, only a few areas are open to visitors, and regulations help keep these areas clean and undisturbed. Extracting oil and minerals could also cause pollution. It already has in the Arctic, when oil from the tanker *Exxon Valdez* polluted the coast of Prince William Sound, Alaska, in 1989. The Antarctic Treaty is an agreement among nations to protect Antarctica, but some people want to turn the continent into a World Park.

The Southern Ocean has already been exploited by humankind. Sealers hunted fur seals almost to extinction, but seal numbers are increasing again. Sealers were followed by whalers who, in the first seventy years of this century, killed tens of thousands of whales and left the blue whale and southern right whale in danger of extinction. Now that it is almost too late, these whales are protected. Fish and krill are now being fished, but international law limits the catch so the birds and mammals that eat them will have enough food.

The polar regions are affected by what is happening many miles (km) away. Air pollution from Russia, Europe, and North America drifts over the Arctic, and, in both polar regions, chemicals called CFCs are destroying the protective ozone layer in the atmosphere. Damaging ultraviolet light shines through this "ozone hole." This kind of pollution can be stopped only by legal action in the countries causing the pollution.

Discarded steel fuel drums litter the Arctic shoreline at Chukotka, Russia.

MOUNTAINS

Snow fields lie on the upper slopes of Africa's Mount Kenya, even though it is only a few miles (km) from the equator.

Mountains develop where sections of Earth's crust called continental plates collide and force rocks upward. The crash of the plates is extremely slow. Plates move only a few inches (cm) per century, and the mountains rise just as slowly. Young mountains are steep and have sharp peaks. Over the centuries, they suffer erosion from the weather and so become shorter and rounded. The European Alps are new mountains, and the Himalayas are still rising, so they have spectacular peaks. The mountains of the British Isles are old and much lower than they once were.

A mountain habitat is like an island in the surrounding habitat. Some species of animals and plants live only on mountains and do not thrive in the surrounding countryside. Because mountains are separated from each other, they often have their own unique species that have evolved in isolation. A species may be found on only one mountain, so it can easily become endangered like animals living on small islands.

Several zones of fauna and flora exist from the base of a mountain to its peak. A mountain rising from tropical rain forest to a height of 19,685 feet (6,000 meters) will show the widest range of habitats. Rain forest grows on the lowest slopes and changes into deciduous forest at about 3,280 feet (1,000 m) and then coniferous forest at 9,840 feet (3,000 m). The trees get smaller with higher elevation until there are only shrubs. This is the level that is often damp and covered in clouds. The forest that grows there is called cloud forest and is found throughout the tropics. The trees are covered with plants that take root on their trunks and branches. They are called epiphytes and include orchids and mosses.

The treeline is that point of altitude on the mountains at which the trees stop growing. Grassland exists above the treeline, and, higher still, only a scattering of very hardy plants, just as on the Arctic tundra. This is called the alpine zone. Some plant species grow in both alpine zones and arctic tundra. They have to survive low temperatures and high winds. Beyond that, only rocks,

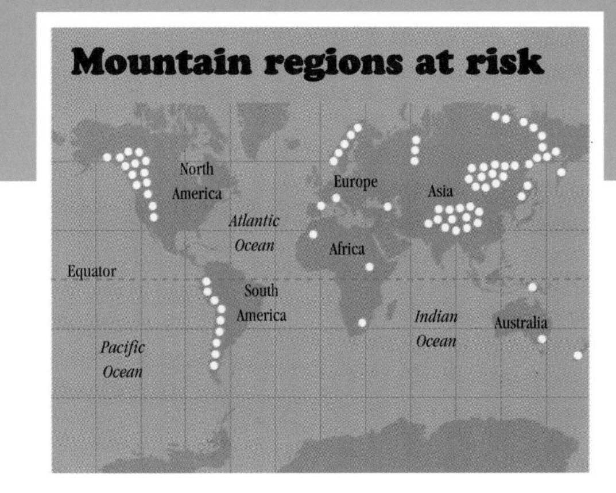
Mountain regions at risk

Despite their clumsy appearance, yaks are very good climbers. Seen here climbing to the Nimaling Plateau in Ladakh, India, they are important livestock for mountain dwellers.

snow, and a few lichens, algae, and insects exist. The height of the different zones depends on the mountain's geographical position. The treeline is above 13,125 feet (4,000 m) at the equator, but below 985 feet (300 m) in northern Scandinavia. The highest forests are at 14,765 feet (4,500 m) in Tibet.

Mountain animals live in a climate as severe as that of polar animals. The mammals are protected by thick fur, and they also take shelter from the cold wind. The chiru, a Himalayan antelope, scrapes a trench to lie in. Some species survive winter by hibernating or by storing food. Others migrate to lower altitudes for the winter. These animals breathe thin air, so they have larger lungs and hearts, and their blood has more oxygen-carrying hemoglobin than that of their lowland relatives.

Most mountain animals are plant-eaters. Small species include the snow voles, marmots, chipmunks, and chinchillas. Large species include the yak, which is a valuable domestic animal in Tibet; the takin, which is related to the Arctic muskox; and the vicuna of the Andes. The ibex, chamois, and Rocky Mountain goat are good at climbing crags and cliff ledges. These animals are preyed on by creatures like the snow leopard, which also lives at high altitudes, or other carnivorous mammals that come up the mountain. Birds can easily fly up and down mountains, but some live at high altitudes. The alpine chough has been seen at 26,905 feet (8,200 m) on Mount Everest, but it comes down to the valleys in winter. The accentors are sparrowlike birds that live on mountains in Europe and Asia, and there are even hummingbirds that live at nearly 16,405 feet (5,000 m) in the Andes. These birds eat insects instead of sipping nectar like most hummingbirds. The Andean condor and the lammergeier are mountain scavengers that feed on the corpses of other animals.

The upper parts of mountains are barren in most cases, but the slopes below the treeline are full of wildlife. They are also the home of one-tenth of the world's human population. An increasing number of people live on mountains, and the natural riches are disappearing. The Incas started to clear the mountain forests of the Andes centuries ago and used the land to graze their herds of llamas and alpacas. The northern Andes are important because they are the home of the wild relatives of potatoes, lima beans, and other crops. Coffee originally came from the mountains of Ethiopia, and wild relatives of maize thrive on the Sierra Manantlan in Mexico. Wild relatives of plants are

Pinnacles of hard rock, like cathedral spires, tower above alpine forests in the Black Hills of South Dakota.

Mountain environments provide habitats for many rare species, such as this Abruzzo chamois goat in the Abruzzo National Park in Italy's Apennine Mountains.

necessary to breed new crop varieties. The rain and snow that falls on mountains is essential for people and wildlife that live on lower ground hundreds or even thousands of miles (km) away. Mountain forests hold the water so rivers flow steadily throughout the year. About half the world's human population lives in southern and eastern Asia. This entire population depends on river water that flows out of the Himalayas and other mountains in central Asia. The water in the Nile River that brings life to the desert in Egypt comes from distant Ethiopian mountains.

In many places, the rise in human population has led to increasing destruction of the mountain habitat. There are now over 70 million people living in the northern Andes, and 90 percent of the forests have disappeared. In the mountains of central Asia, 7,720 square miles (20,000 sq. km) of juniper forest have been reduced to 1,775 square miles (4,600 sq. km). Without the tree cover, the soil is washed away, and the ground is eroded. This leads to mud slides and floods that cause enormous damage. In places like Nepal, farmers cultivate steep slopes by making terraces that work to preserve the soil.

As well as cutting trees for timber and fuel or converting the forests into farmland, mountain people earn money by growing plants that yield drugs. In one year alone, 27,180 acres (11,000 hectares) of montane forest in Colombia were turned into poppy fields for making opium.

Winter sports are also causing problems. Millions of people come into the mountains for skiing. New roads and hotels are built, and forests are thinned out or cut down to improve ski runs. The result is increased erosion. Without the protection that forests provide, villages in nearby valleys are in danger from avalanches. Even a vehicle parking area is a problem when rain and melting snow pour off it.

Humans now realize the importance of the mountain habitat for providing healthy water supplies and preventing floods, and mountain areas are being protected to save the plant cover. Governments of countries with mountains are trying to find ways for the inhabitants of mountain areas to improve the quality of their lives without harming the habitat that is vital for the entire nation.

Leisure developments, such as this ski resort under construction in the Rhodope Mountains in Bulgaria, combine with overgrazing to threaten the delicate balance of the alpine habitat.

CONIFEROUS FORESTS

New forest growth in Finland, where timber and pulp are major products for export.

An immense belt of coniferous forest runs through the northern sections of North America, Europe, and Asia. It lies between the treeless tundra to the north, and the temperate forests and open grasslands to the south. The coniferous forest has a cold climate and is often called the boreal forest, after Boreas, the ancient Greek god of the north wind. Summers in the boreal forest are short, and there may be only two months in the year that are free of frost. Temperatures in the Siberian forests go as low as -76°Fahrenheit (-60° Centigrade) in the winter. Another name for the northern coniferous, or boreal, forest is *taiga*, which is the Russian word for a marshy pine forest. The forest soil is often poorly drained. Melting snow and rain does not drain properly, so pools and bogs form.

These forests can only be inhabited by hardy species because of the cold climate. There is a little moisture in the form of rain or snow, but this moisture is frozen for most of the year. The dominant trees of the boreal forest are coniferous trees, such as spruce, pine, fir, and hemlock, and a few broad-leaved deciduous trees, such as willow and birch. Coniferous trees are adapted to cold, dry conditions. The needle-shaped leaves have a thick, waxy covering that stops them from losing water, and the conical shape of the trees sheds the snow that would otherwise accumulate and break the boughs. The evergreen habitat lets them photosynthesize and start growing as soon as there is sufficient warmth and moisture in spring. Deciduous trees do not grow well in these conditions because they have to grow new leaves every spring and, with short summers, they lose more food material when shedding their leaves in autumn than they gather during the brief summer growth. The trees of

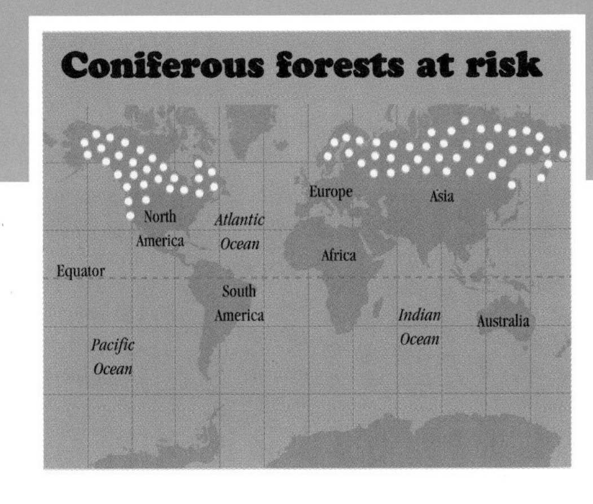

the boreal forests grow so densely that they block out the light at ground level, and very little can grow under the trees. There is not much undergrowth except for a layer of lichens, ferns, and low-growing shrubs, such as dwarf birch, bilberry, cranberry, and bearberry. The berries are important food for animals that gorge on berries in autumn to fatten up for the winter.

An important feature of the coniferous forest is the uniformity of plants and animals. Instead of the diversity that characterizes tropical rain forests, the same species can be found around the world in the forests of Alaska, Scandinavia, and Siberia. This is because these forests are a fairly new habitat that became occupied only when Ice Age glaciers retreated a few thousand years ago. There has not been time for new species to evolve.

Winter is a very hard time for animals and birds in coniferous forests. There is plenty of insect life in summer, especially mosquitoes and blackflies that make life miserable for the animals whose blood they suck. The adult insects die in winter, but eggs and larvae survive. Many of the birds that eat insects migrate south for the winter, but the seed- and berry-eaters remain behind. These include nutcrackers, jays, crossbills, and capercaillies. The nutcrackers and jays store food and find it again even when their hoards are covered with snow. Crossbills have beaks with crossed tips for splitting cones to extract the seeds. The capercaillies are large game birds that eat conifer buds and shoots.

Mammals inhabiting coniferous forests include large reindeer, caribou, elk, and moose, and small squirrels, voles, and lemmings. These mammals are the prey of wolves, wolverines, stoats, and lynxes. The beaver is an important animal because it dams streams that flood the surrounding country to create new habitats for a variety of animals and plants. These animals survive the winter by conserving their food supplies. They spend much of their time resting. Bears sleep in dens for long periods of time, but they are not true hibernators. Capercaillies dig holes in the snow and emerge only to feed, but the voles and lemmings remain active in tunnels under the snow.

There is no equivalent of the boreal forest elsewhere in the world, although there are other forests with conifer, or cone-bearing, trees, such as monkey puzzle trees in Chile, Parana pines in Brazil, yellowwoods in South Africa, and kauri pines in New Zealand. Only a remnant of the famous Cedars of Lebanon survives.

The traditional human population of the boreal forest lived as hunters or, as in northern Europe and Siberia, reindeer herders like the Sami (Lapps). The outside

A capercaillie, frequently hunted as a game bird, "displays" in a tree.

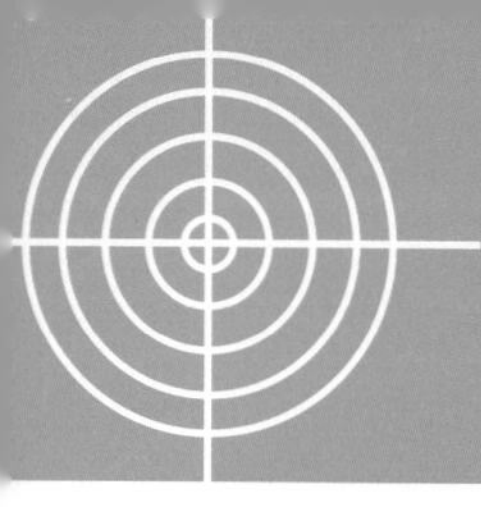

The pine marten, or sable, has a coat of fine, warm fur that has caused it to be hunted extensively by fur trappers.

world became interested in the forest because of the valuable furs that could be taken from sables, martens, and lynxes. The trees were also cut for timber and resin. The coniferous forests are so extensive that this had little effect until recently. Although the forests are vast, remote wildernesses, new roads and railways make access easier. Trees are now a source of pulp for making paper and construction lumber, as well as timber. Clear-cutting large areas completely ruins the habitat and leads to erosion. Even less intensive logging causes problems for animals, such as woodpeckers, that require old trees. The woodpeckers, which feed on insects living in wood, need holes for nesting. These are found only in mature trees, so woodpeckers cannot inhabit areas that have been logged.

The coniferous forest spreads down the western side of North America as far south as California. The main trees are Douglas fir, hemlocks, Sitka spruce, and red cedar. The Douglas fir lives for over one thousand years and grows to 295 feet (90 m) in height. These forests, like large areas of the Russian taiga, have been cut for timber. Only 10 to 20 percent of the original forest remains. The old forest has a much richer wildlife than forest that has regrown after cutting. The northern spotted owl is a rare owl that lives only in old coniferous forests, and conservationists are trying to prevent logging to save the owl and other forest wildlife. An old-growth coniferous forest, the Olympic Rain Forest in the state of Washington, is preserved as a national park.

Coniferous forests are also threatened by acid rain. Chemicals released into the air from burning coal and oil in power stations and factories react with the moisture in the air to make sulfuric acid. Nitrogen oxides from industry and motor vehicles similarly react with atmospheric moisture to make nitric acid. These acids pollute the environment and cause what is known as acid rain. Scientists have known for a long time that acid rain kills lichens and mosses growing on tree trunks and branches, but it is now killing the trees as well. Acids from factories in industrialized countries also drift on the wind over the coniferous forests and cause severe damage. Industrialized countries are trying to cut down the amount of oxides polluting the air, but this is very difficult. In Sweden, thousands of tons of lime have been spread over forests to neutralize acid rain. It is possible, however, that the lime may also harm the forests in another way.

A forest fire can cause complete devastation to plant and animal life through the destruction of habitats, even though the forest itself may recover within a few years.

TEMPERATE FORESTS

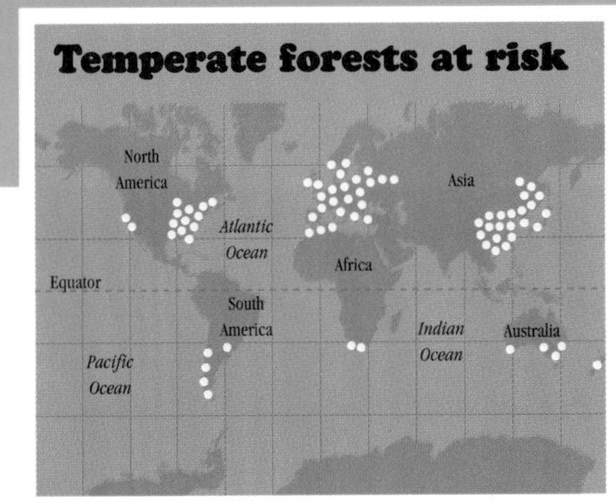

Outside the tropics, the type of forest that grows in a specific geographical region depends on changes in temperature throughout the year. Near the Arctic, where the winters are long, there is the coniferous forest, or taiga (page 18). South of this region is a belt of broad-leaved forests of deciduous trees that lose their leaves in winter. A temperate climate has plenty of rain, and winters are mild, but they are still cold enough to prevent the trees from growing. The leaves of these deciduous trees could be killed by frost, so they shed in autumn. Nutrients are reabsorbed into the trees, so the leaves die before they fall. The leaves rot on the forest floor, and their remaining nutrients form a rich soil.

Deciduous forest was the original habitat over much of Europe, eastern Asia, eastern North America, and southern South America. Compared with rain forests, they contain a small number of tree species, including oak, lime, ash, hickory, beech, and maple. They grow to 98 to 130 feet (30 to 40 m) and live for three hundred years or more. Plant life in a deciduous forest depends on the type of soil and climate. For instance, beech and ash grow in dry, chalky soil, and oak grows in wet clay. A layer of shrubs or small trees such as hazel, hawthorn, and holly live under the primary trees. Because there are no leaves in early spring, plenty of sunlight reaches the ground, and grasses and herbs grow under the trees. Carpets of bluebells, violets, anemones, and other plants flower and die before the tree leaves grow.

The trees grow new leaves and their flowers bloom when the temperatures rise in spring. The flowers of oak, beech, and elm are tiny, and they are fertilized by pollen that is carried along by the wind. Maple, lime, and chestnut have large flowers that are pollinated by insects. The seeds and fruit are an important food source for animals when they ripen. Many birds and mammals eat nuts, acorns, chestnuts, and holly berries. They are also stored by squirrels, mice, jays, woodpeckers, and nutcrackers to eat in winter. More berries, such as strawberries, and an autumn crop of fungi for the animals to eat also grow on the forest floor.

The buds and leaves of forest plants are food for many animals. Deer browse on tree leaves and graze on grass and herbs growing on the forest floor, but insects are more numerous than mammals. A single oak tree can have fifty thousand caterpillars chewing its leaves and many times more aphids sucking the leaves' juices.

A pair of young, inquisitive gray squirrels peers cautiously from their nest in the hollow of a plane tree. Tree hollows in mature trees provide shelter for many woodland animals.

As winter approaches, the leaves on deciduous trees turn to rich golds, yellows, and browns before they fall to the forest floor.

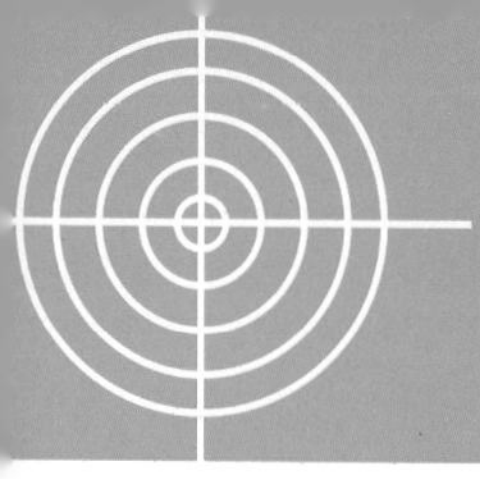

Attacks by these insects can be so severe that the trees lose all their leaves. Many birds, such as titmice and treecreepers, and mammals, such as bats and shrews, eat these insects. Earthworms inhabit the rich soil beneath the trees. They help make the soil fertile and are a food source for moles, badgers, and birds.

Hardly any temperate deciduous forests remain in their original condition. The rich forest soil is good for growing crops. Forests in Europe and Asia have been cleared and turned into farmland ever since prehistoric times. Stone Age farmers began to grow crops in Europe thousands of years ago. In their own country, Native Americans lived in the forests and hunted animals and gathered plants. There was very little change until European settlers cleared the forests for agriculture.

Coppicing trees is one method of harvesting wood from deciduous woodlands.

Unlike the poor soil under tropical forests (page 26), the soil that formed under deciduous forests can be used for growing crops for hundreds of years. Very small patches of natural forest have survived in remote ravines where it was never possible to farm, and there are a few larger patches in eastern Europe that were once hunting preserves for the nobility. The largest of these is the Bialowieza National Park, which was an ancient hunting park for Polish kings. There are only 19 square miles (50 sq. km) of natural forest in the park, but there are eighty-one species of trees. Some of the oaks and ashes have grown to 148 feet (45 m). The forest is the home of over two hundred bird species as well as wolves, lynxes, wild boar, and elk. The European bison, or wisent, became nearly extinct in World War I, but it has been reintroduced through captive breeding programs.

The small patches of temperate forest that survive are different from the original forest due to the actions of humans. Cutting trees for timber and firewood does not permanently destroy the forest because the trees grow again, but it causes changes in the type and shape of the new trees and other plants. One way to harvest a forest is to coppice the trees. This means cutting them down every few years and letting them regrow from the stumps.

Plantations of a single species of tree sometimes replace natural and altered forest. A foreign species often produces better timber than the natural trees, but it does not provide as good a habitat for the native wildlife. In Sweden, the middle-spotted woodpecker has become extinct, and the white-backed woodpecker is rare because cutting trees for timber causes changes in its forest home. These birds need old trees that provide insects to eat and holes for nesting. In North America,

A carpet of bluebells thrives in the rich soil of the woodland floor in this coppiced wood in Dorset, England.

large areas of forest that had been cleared for agriculture have been allowed to regrow. The new woods look natural, but they are not the same as the original forest.

Forests are often in great demand for use as farmland, so these habitats will continue to disappear unless they are preserved in a nature reserve. Even so, forests may continue to change and disappear unless they are carefully managed. A natural forest is always changing. Fire started by lightning burns the vegetation but stimulates new growth, and even one tree falling lets sunlight in for herbs and shrubs to grow. One forest in eastern Canada that had remained undisturbed for two hundred years was home to fewer species of animals and plants than neighboring areas that had been cut or burned. Forest management may have to control some animals, too. An abundance of deer feeding on undergrowth and young trees, for instance, may devastate the forest, leaving only old trees to survive. When the old trees die, there will be no forest left. The same thing happens when sheep are allowed to graze in woods.

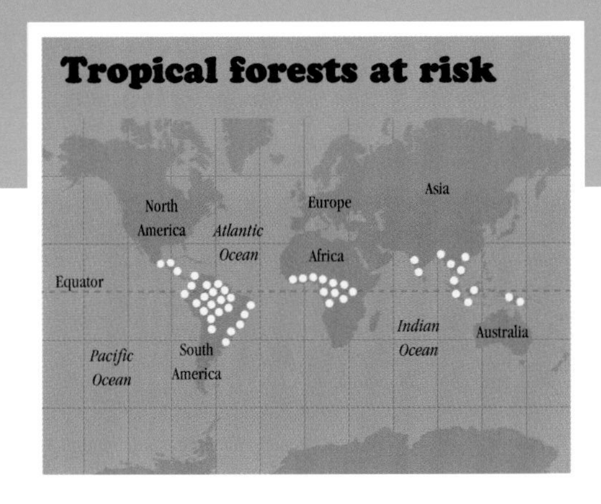
Tropical forests at risk

Rain forests grow where there is an average yearly rainfall of 6.5 feet (2 m) or more. Most rain forests are in Earth's tropical areas, but some do exist in temperate countries. Tropical rain forests are evergreen because they receive rain throughout the year. Tropical dry forests are deciduous. These forests grow to the north and south of the rain forests and receive large amounts of rain for only part of the year. This is often during a rainy season caused by the monsoon, a wind that brings rain from the sea. The trees shed their leaves during the dry part of the year.

A tropical rain forest is often called a "jungle." There are two kinds: lowland and montane. Lowland forests are the most common, and they contain several layers of vegetation. The tallest trees, which grow from 98 feet (30 m) to as much as 197 feet (60 m) high, form the canopy, or roof, of the forest. These trees have tall, straight trunks and broad crowns of foliage. Below the canopy is another layer of trees with a roof at 50 to 60 feet (15 to 18 m). Nearer the ground is a third layer with a roof at 20 to 50 feet (6 to 15 m). The roofs prevent light from reaching the ground, so very few plants live at ground level. But climbers and epiphytes (plants that grow on other plants) still grow high in the trees. If one of the large trees blows over, sunlight shines through the gap, and plants immediately start to grow. More low growth also exists on the banks of rivers or the sides of roads where light can get in. Montane rain forests grow on high ground, between 2,950 and 10,500 feet (900 and 3,200 m). The climate is cooler and the rainfall less, so trees are smaller than in lowland forests.

Rain forests are important because they provide homes for a very large number of animal and plant species. Approximately 100,000 plant species live in the world's tropical rain forests, with as many as several hundred species of trees alone in one-third square mile (1 sq. km). Hundreds of insect species can live on a single tree. Most of the larger animals, such as monkeys and apes, big cats, bats, sloths, squirrels, and many birds, are good climbers. The fruits and leaves of the trees also provide plenty of food. Relatively few animals live on the forest floor, but there are reptiles, amphibians, insects, and various large animals.

Hot, humid conditions encourage spectacular plant growth in tropical rain forests like this one on the Caribbean island of Guadeloupe.

Many rain forest plants have important commercial value. These include mahogany, teak, and other trees with valuable wood, the rubber tree, banana, coffee, cocoa, ginger, nutmeg, and many more that produce substances used in food and medicine. The cinchona tree provides quinine, a drug used to prevent malaria. Some plants, like rubber and cinchona, can grow in plantations, but many timber trees are taken from the wild forest.

Rain forests help control the weather. Half the rain falling on the Amazon rain forest returns to the clouds and falls again. This recycling helps keep the forest wet. If part of a rain forest is cut down, the amount of rainfall decreases. There is no "sponge" of vegetation to hold the water, so rivers start to dry up or flood rapidly when it rains. When the trees are cut down, erosion increases, and the loss of soil per year rises to over 110 tons (100 tonnes) from 66 pounds (30 kg). Destroying a rain forest affects the climate many thousands of miles (km) away because the forest affects air circulation.

Small numbers of people have lived in rain forests for thousands of years. They have little effect on the forest because they cut down very few trees. They also plant trees and grow their crops between existing trees. This

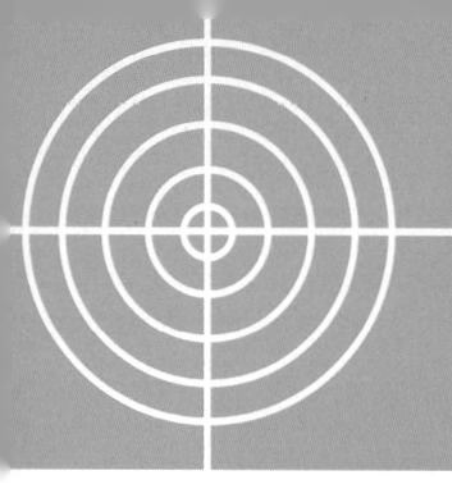

This colorful red-eyed leaf frog, perched on a heliconia plant, lives in the Costa Rican rain forest in Central America.

In Amazonia, Brazil, and other parts of the world, tropical rain forest is being cleared on a massive scale with no regard for the preservation of precious plant and animal life.

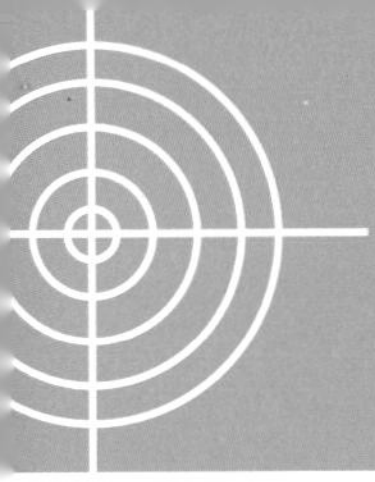

type of agriculture only works if the number of farmers is small and if they understand how to live wisely in the forest. Today, millions of people are moving into the forests and destroying them. Timber companies use machinery to cut and remove trees. As well as the highly valuable trees, such as mahogany, other less valuable trees are cut and used for chipboard, plywood, and other products. In many places, the forests are completely destroyed to make plantations of rubber, tea, coffee, tobacco, and other crops. Some forests have been turned into grassland for cattle ranching, and large areas of forest in Brazil now grow sugarcane to make fuel for motor vehicles.

Settlers who clear small patches of forest for growing crops have caused the greatest destruction. Most of the nutrients in a rain forest are stored in the trees. When the trees are cut down, only poor soil remains. After two or three years, the ground becomes infertile, and the farmers must clear new areas of forest. This is called shifting cultivation, and it destroys huge areas of rain forest around the world. When the cleared ground is abandoned, the forest regrows as bushland or secondary forest. A complete return to rain forest takes two centuries or more. The human population in Madagascar, for example, grew from 5.4 million people in 1960 to 12 million in 1990. Most of these people live by farming. As the human population increases, there is less land available for each family and more pressure to grow crops. Because the land is not allowed to lie fallow and recover its fertility, it eventually becomes too poor for farming. More forest has to be cut down, and there will soon be very little left.

When a rain forest is cut down or burned, nearly all the plants and animals disappear because they need the rain forest habitat to survive. Rain forest species often live only in a small area, and they become extinct as soon as their home is destroyed. Birds such as Bannerman's turaco and the banded wattle-eye are now under threat as their habitats disappear. No one knows exactly how many species of animals and plants live in the rain forests or how fast they are becoming extinct as the forests disappear. It is very important to save these species and their forest homes. The forests protect the land and climate. At the same time, they provide animals and plants for use as food, medicine, and other products.

Only a little more than half the world's rain forests remain. They are being destroyed at the rate of 25 million acres (10 million hectares) a year. About 4 percent of the remaining rain forests are in reserves, but this does not always prevent destruction. The rain forests will be saved if they are managed properly. This means careful harvesting of trees and other species so the forest is not devastated.

New forest is being cultivated in the Wotu region of Sulawesi in Indonesia, but it can take many years for a tropical rain forest to reach maturity.

GRASSLANDS

Grassland is a habitat that can cover many square miles (sq. km) with various species of grasses as its main plants. There are 10,000 species of grass that vary in height from a few inches (cm) to several feet (m). Bamboos are a kind of grass, and some grow to 98 feet (30 m). Unlike other plants, grasses grow from the bottom of the stem, so they continue to grow even when the top of the stem has been eaten by an animal or cut by a lawnmower. A variety of low plants called herbs grow among the grasses.

Although grasslands may also have some scattered trees, grassland forms where forests cannot grow. This may happen because of insufficient rain to support many trees. Grasslands usually grow where rainfall is less than 47 inches (120 cm) a year, and most of this falls during short, rainy seasons. Also, any seedling trees that try to grow are likely to be eaten by animals that graze on the grassland. When the disease known as rinderpest killed huge numbers of animals in East Africa in the 1880s, woodland soon began to grow over the grasslands. As the animals' numbers increased again, the new trees were eaten and the grassland returned. A third reason for an absence of trees is natural fires that are started by lightning. Grasses and other herbs quickly grow back because their seeds and roots survive underground, but the fires kill the trees. Some grasslands are created by people who deliberately start fires to burn the trees and bushes. This creates more grazing area for their domestic animals.

Large grassland regions exist in both the tropical and temperate regions of the world, especially in the center of continents with low rainfall. Tropical grasslands grow where the annual rainfall is between 30 and 60 inches (75 and 150 cm) a year. These areas include the savannas of South America, Africa, southern Asia, and northern Australia. The African and South American savannas lie to the north and south of the tropical rain forests. Temperate grasslands have a lower rainfall of 10 to 20 inches (25 to 50 cm). These areas include the prairies of North America, the steppes of eastern Europe

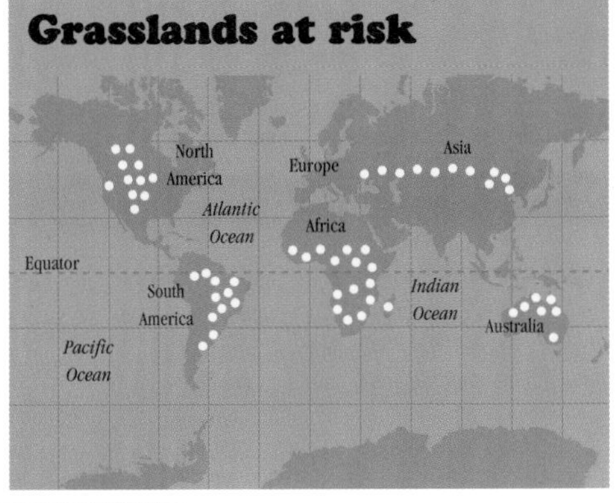

Grevy's zebra, which is an endangered species, is a typical grassland grazer.

and Asia, the pampas of South America, the South African veld, and the downlands of Australia.

Grasslands are home to a large number of animals that eat only grass and other short plants. Grass-eating animals are called grazers. Grass stems are very tough and would wear down even normal teeth. Grazers have teeth that grow continuously, so they renew themselves as fast as they wear down. Grass is also difficult to digest, and many grazers have bacteria and other microscopic organisms in their stomachs and intestines that help with digestion.

The African savannas are famous for herds of large animals. Zebras, elephants, rhinoceroses, buffaloes, and many kinds of antelopes, such as wildebeests, gazelles, eland, and impala share the same country. They can live together because they have slightly different feeding habits. The herds move around the savannas in search

This view across the Garamba National Park in Zaïre shows a typical grassland area containing just a few trees.

of the best grasses to eat, and they are hunted by large carnivores such as lions, leopards, cheetahs, wild dogs, and hyenas. There are also many small grazing mammals that include hares, rats, squirrels, and birds.

The North American prairies were once inhabited by 60 million bison. These animals lived in large herds that traveled southward in winter to escape frost and snow and returned northward to find fresh grass growing where the snow had melted. By eating seedling trees, they helped maintain the grassland. They also shared the prairies with 50 million pronghorns and vast colonies of prairie dogs. The equivalent animals on the European and Asian steppes are wild horses, an antelope called the saiga, and many kinds of small mammals, such as marmots, susliks, pikas, gerbils, and voles. Australian grasslands are inhabited by herds of kangaroos. But insects sometimes have greater impact than the much larger animals because of their high populations. The weight of termites on the African savannas, for instance, is at least twice as much as the herds of antelopes and

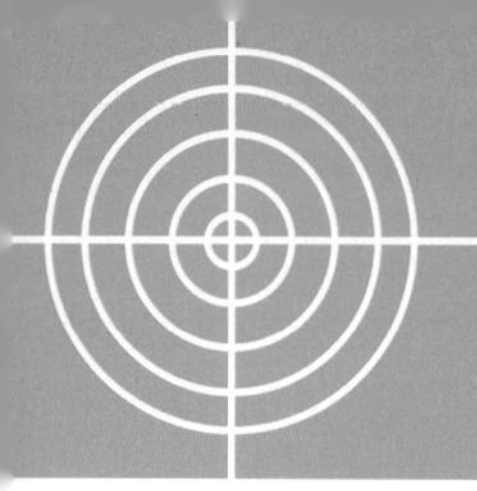

Vehicle tracks scar the surface of this degraded savanna on a plain near Anivorano in northern Madagascar.

zebras. Grasshoppers and locusts sometimes form swarms that move across the grassland and eat all the plants.

At one time, nearly half the world's land surface was grassland. About half has disappeared through human activities. At first, people increased the area of grassland by destroying trees with fire or letting their sheep and goats eat the seedlings, but grasslands later declined through other human activities. The British Isles has no natural grassland, and the large areas of grassland that now exist were created by domestic animals. In the last 100 to 200 years, an enormous amount of grassland has disappeared. The American prairies and Asian steppes have been plowed to become the world's largest food-producing areas. The original grasses and herbs have been replaced by wheat, cotton, and other crops. In other places, grasslands have been used to farm rice and vegetables.

Even where the grassland has not been plowed up for crops, the original habitat has been destroyed because cattle, sheep, and goats have replaced the native animals. Too many animals destroy the vegetation and cause erosion. Humans also set fire to the vegetation, but their fires damage the grassland instead of helping it. The fires are too big and too frequent, so the vegetation does not get a chance to recover.

Because grasslands are so valuable for agriculture, it is not easy to preserve them. The American prairies and the Asian steppes have almost disappeared. Only a few small reserves with the original plants and animals remain. In Australia, European settlers plowed the native grasslands for crops and raised sheep and cattle on them. When national parks were established to conserve wildlife, most of the grassland was already gone. In Africa, the increasing human population needs the savannas for crops and domestic animals. So there is less room for herds of antelopes and other spectacular animals except in national parks. Even these protected places are threatened, however, because local people use them for their domestic animals. The Serengeti Park in Tanzania is one of the most famous places in Africa for wildlife. Yet it is not safe for the native animals because the Masai tribespeople want to move their herds of cattle into the area.

This unusual construction is a termite spire. Although insects may be less conspicuous than large grazing animals, they are just as valuable to the ecology of grasslands.

DESERTS

Deserts are places with a special community of animals and plants adapted for life in dry conditions. They are also home to more than 10 percent of the world's human population. The definition of a true desert is a dry place where rainfall is less than 9.8 inches (250 mm) per year. Semideserts are places where the rainfall is up to 20 inches (500 mm). The driest deserts may not have any rain for years. The Namib Desert of southwestern Africa receives its moisture as fog that comes in from the sea.

Deserts cover nearly one-third of the world's total land surface. Most of them, like the Sahara Desert, the Arabian Desert, the Sonora Desert of California, and the Australian deserts are very hot, but a few deserts are cold. The Gobi Desert of central Asia, for example, is very cold in winter, and the land surrounding the Arctic Ocean is a cold desert. Hot deserts have an average summer temperature above 86° F (30° C), and cold deserts have an average winter temperature below 32° F (0° C).

Deserts occur where there is a natural lack of water. There is very little rain or other moisture, and the ground dries quickly through evaporation. A belt of deserts exists around the world on each side of the equator to the north and south of the tropical rain forests. They develop as a result of convection in the atmosphere. The sun heats moisture-bearing air at the equator that rises through the atmosphere. The air cools as it rises, and the moisture falls as rain. The dry air then moves toward the poles and sinks over the desert belts. It becomes drier and prevents wetter air from moving in.

Some deserts, such as China's Taklamakan Desert, are dry because they are in the center of a continent. The air reaching them from the ocean has already lost its moisture by falling as rain before it reaches them. Others are cut off from the ocean by mountain ranges. The

Although the sands of the Temet Dunes in Niger look inhospitable, many animals are specially adapted for life in hot, dry deserts.

mountains trap the rain, so the air reaching the desert on other side is dry. This is called a rain shadow. The Mojave Desert is sheltered from the Pacific Ocean by the Rocky Mountains.

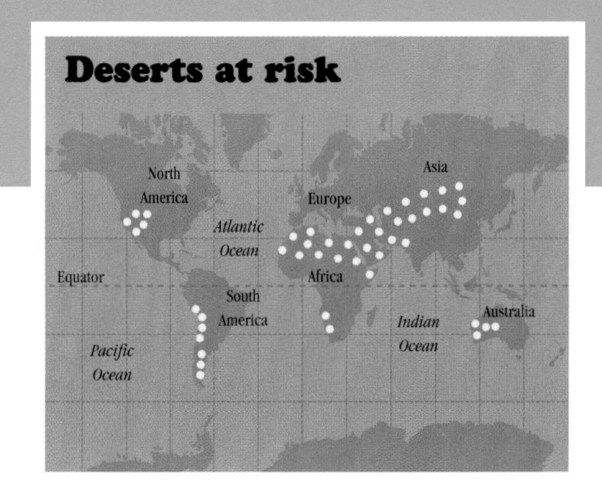

Every desert has places, called oases, that have plenty of wildlife. These oases are fed by spring water and rivers that flow through the desert. The rivers may be dry wadis that fill with water only after rainstorms, but enough moisture remains in the soil for plants to keep growing. Desert plants have several ways of saving water. Spiny leaves and thick, waxy cuticles covering their surfaces reduce water loss. Some plants have roots that spread under the surface of the soil to absorb as much water as possible from rain showers or dew. Others have very long roots to reach water many feet (m) deep. Cacti have fleshy stems that act as water stores. When there is some rain, desert plants flower and seed very quickly. Seeds that have been lying in the dry soil sprout as soon as they are wet. Some seeds wait over one hundred years before sprouting.

An agave in bloom at Puebla, Mexico, is surrounded by cacti. The agave species are succulents that store moisture in their tough, spiky leaves. Cacti use their large, fleshy stems to store water.

Desert animals try to avoid heat. Small animals spend the day in burrows, where it is cool and humid, and they only come out at night. The African ground squirrel comes out during the day and holds its bushy tail over its back as a sun shade. Desert hares use their long ears as radiators to lose heat. Thick coats of fur or feathers are also a good protection. A camel's hair can reach 158°F (70°C), while its body temperature is only 104°F (40°C). Camels drink as much as one quarter of their body weight in a few minutes. This amount lasts for several days. Plant-eating small mammals, such as gerbils and jerboas, can live on dry food without drinking, but they sip dew from plants when they can. Carnivores get water from the blood of their prey.

A pair of musk oxen gallop over the snow-covered tundra of Victoria Island in Canada's Northwest Territories. Deserts occur in places of extreme cold or intense heat.

Humans live mainly on the edges of deserts, where up to 7.8 inches (200 mm) of rain fall per year, especially at spring-fed oases. They grow a few crops, and animals, such as camels, goats, and sheep, graze on desert vegetation. Desert people are often nomads, such as the Bedouin, who travel in search of the best vegetation for their livestock. Others are farmers who irrigate their fields with water from rivers or wells.

Traditional desert inhabitants did not overexploit their environment, but many more people in modern times have moved into the deserts, and fragile desert environments are suffering. Pumping water from underground stores and rivers has irrigated large areas of desert for farming. This causes the natural plants and animals to lose their habitats. Water supplies are drained, so the desert eventually becomes drier. When irrigation water evaporates, salt stays behind in the soil.

Duweltjies flower after a rainfall in South Africa's Kalahari Desert. Many plant species are adapted to react quickly to the arrival of rain in order to reproduce successfully.

This is called salinization. Plants cannot grow in this salty soil. The increased use of the desert fringes also leads to erosion. Where natural vegetation has been destroyed, sudden rainstorms wash the soil away and carve gulleys in the ground. Agriculture becomes impossible, and wildlife cannot survive. This process is called desertification. It creates lifeless, false deserts at a rate of 49 million acres (20 million ha) a year. A long-standing problem, desertification caused the ancient civilization of Mesopotamia to collapse thousands of years ago.

Oil has made desert nations extremely rich, but it has also led to desert animals being overhunted (*see* Arabian oryx, *Endangered Mammals!*). Destruction caused by vehicle wheels and pollution also affect the desert habitat. The worst example of pollution was the Gulf War of 1990-1991, which left lakes of oil and churned-up soil in the desert of Kuwait. People also come into deserts for recreation. In California deserts, hundreds of thousands of motorbikes, dune buggies, and other motor vehicles invade the terrain. Their wheels tear up the vegetation and crush tortoises in their burrows.

The process of desertification is turning the natural desert habitat into false desert. Although life in deserts is difficult, modern technology has allowed large numbers of people to move into drier parts of the world. Desert wildlife then becomes dependent on the creation of large reserves, where hunting and environmental damage are forbidden. Where desert dwellers try to make a living from the desert, their way of life must be sustained in a manner that will preserve their own environment. This will be difficult to achieve because the human population continues to increase.

WETLANDS

Wetlands is the name given to the wet areas on the world's landmasses. They include lakes and rivers; waterlogged ground, such as marshes, swamps, and bogs; and floodplains, which are areas along rivers and lakes that flood regularly. The Pantanal is a floodplain in southwestern Brazil where the Amazon River floods 77,200 square miles (200,000 sq. km) of forest every year. During the floods, it is inhabited by all sorts of fish and even dolphins.

Wetlands are stages in the hydrological cycle. Water evaporates from the sea, and moist air blows over the land where clouds form and drop their moisture as rain or snow. The water then soaks into the soil, gathering especially in wetlands and flowing in rivers back to the sea. Life on land would be impossible without this cycle.

The importance of wetlands for animals and plants, and also for humans, is sometimes forgotten. There are still over 2 million square miles (5 million sq. km) of wetlands in the world, although vast areas have been drained. The United States has lost over half its wetlands, and the United Kingdom and the Netherlands have drained 60 percent of theirs. Over 90 percent of New Zealand's wetlands has been lost since Europeans settled there two hundred years ago. However, some new wetlands have been created as reservoirs and gravel pits, although these are not as good for wildlife as natural wetlands.

The Kafue River winds its way through southeastern Zambia's Kafue Flats in Africa. The Kafue Flats extend almost 125 miles (200 km) on either side of the river.

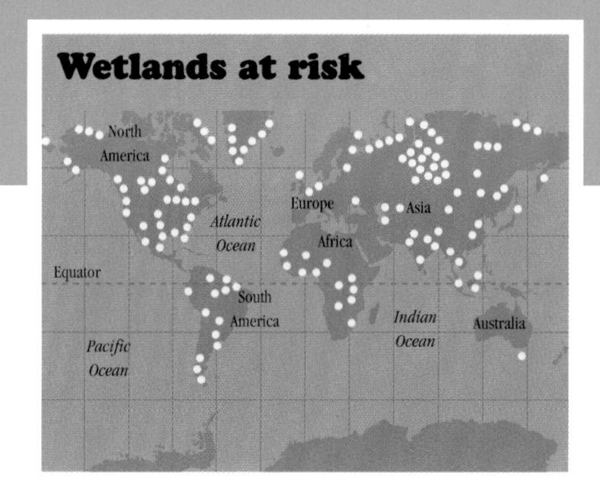

Wetlands are always changing. Rivers change course, washing away land but also creating new land. Rivers also change from their source to their mouth. Many rivers start in the mountains, where they are fast flowing. Any plants and animals inhabiting the rivers there must be strong enough to avoid being washed away. When the river flows over flat country beyond the mountains, the water moves slowly, allowing more plants to grow in it. The plants attract more animals. The river becomes tidal as it nears the sea, with the rising tide driving seawater up the river. Fewer animals can live in tidal waters because the waters are too salty.

Many lakes and ponds change very slowly, but they do eventually fill in with land and disappear. The slow change to dry land that occurs is called an ecological succession. Sediments from the surrounding countryside are brought in by rivers and streams, and water plants die and add organic material to the bottom. As the sediments accumulate, more plants take root in the shallow water. The lake or pond slowly turns into a swamp or marsh and then into dry land so that different types of plant life, even trees, will take root.

The varied wetland habitat provides homes for many species of animals. In East Africa, Lake Tanganyika has 214 species of fish, four-fifths of which cannot be found anywhere else. As well as being the home of water animals, wetlands are important for land animals as places to eat and drink. Kafue Flats in Zambia is the floodplain of the Kafue River, a tributary of the Zambezi River. When flooded, it becomes the feeding ground for over 400 species of birds. Then the floodplain dries out, and a crop of grass feeds herds of zebras, wildebeests, and antelopes. Wetlands are also staging posts for migratory birds. For example, one hundred million ducks and geese migrate through North America and Mexico. These animals depend on lakes, ponds, rivers, and marshes to rest and feed. Unfortunately, these valuable wetlands have been disappearing over the last sixty years.

Wading birds, such as this black-headed heron in Kenya's Nairobi National Park, have adapted to the wetland habitat. They have long legs, and their wide feet allow them a firm footing on the soft mud beneath the water.

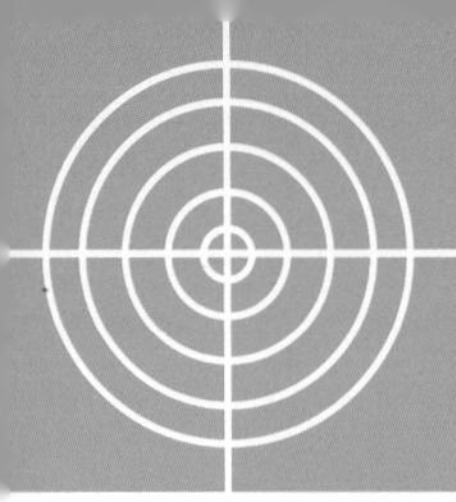

Siberia's Lake Baikal contains the largest volume of fresh water in the world. But the species population is being affected by local deforestation and industrial pollution.

Throughout history, many people have thought of wetlands as wastelands with little or no value. Swamps that once covered thousands of square miles (sq. km) have been drained for agriculture. People have even drained ponds or filled wet corners of fields to get a little more land for cultivation. Rivers have been polluted by industry and agriculture. But wetlands play a vital part in nature, supporting a vast amount of plant and animal life.

Despite their importance, wetland areas are still threatened by existing or new development. Few major rivers in the world have not been dammed to provide water supplies for cities, irrigation for crops, or hydroelectric power. These may benefit some, but the wetland habitat is damaged, and plants, animals, and local people suffer. The Aswan Dam on the Nile River in Egypt was built to control flooding and to provide hydroelectric power. But the annual flood made the land on each side of the river very fertile. Now the desert is spreading over the farmland, while the silt that created the original fertility is accumulating behind the dam.

Irrigation is another threat; it increases the amount of crops that can be grown, but it can be disastrous for wetlands. The Aral Sea was once the world's fourth largest lake. Since 1960, the rivers that fed it have had so much water taken out for irrigation that the Aral Sea is shrinking. Thousands of tons of fish used to be caught every year, but the fishing industry has been destroyed.

At one time, rivers were straightened to let water run quickly to the sea to prevent floods. Now scientists know that a good way to prevent serious floods is to allow wetlands to hold surplus water. Poyang Lake in China

stores one-third of the annual floodwater entering the Yangtze River and reduces flooding by half. So, in some parts of the world, straightened rivers are being returned to their original courses.

Wetlands have always been used as convenient places to dispose of waste, and many wetlands are becoming severely polluted. Many factories and mines get rid of poisonous substances by pouring them into rivers and lakes, and pesticides used on farmland wash into the wetlands. When large amounts of sewage are poured into a river or lake, microorganisms and algae grow on it. These use up the oxygen in the water, so fish and plants cannot survive. Over forty major rivers in Malaysia have been described as "biologically dead." Siberia's Lake Baikal has the largest volume of freshwater in the world. It is home to some unique species, such as the Baikal seal, but local industries pour pollutants into the lake, and cutting the surrounding forests causes erosion. Acid rain, caused by airborne pollutants from industrial output, is another problem for wetlands.

Biologists have shown that wetlands are very productive and that they support large populations of animals and humans. Many people around the world rely on fish they catch in rivers and lakes or on cattle that feed on the surrounding wetlands. Crops such as sago that grow in wetlands give better yields of food than crops grown on the land after it has been drained. Wetlands also play an important part in limiting flooding and providing a steady flow of clean water to the surrounding country. They are, in effect, giant sponges that soak up and slowly release water.

The significance of wetlands is more and more widely appreciated. Some of the most important remaining wetlands are protected by the Ramsar Convention, named after a town in Iran. This is an international agreement for turning wetlands into reserves for wildlife and for emphasizing their importance in providing water supplies.

Queen Victoria water lilies grow in ponds beside the Rio Negro near Manaus in Brazil. The Rio Negro flows into the Amazon River at Manaus.

COASTS

Around every continent and island is a narrow strip called "coast" that serves as a meeting place of the sea and land. Coastlines can have many different characteristics, including rocky and sandy beaches, estuaries, salt marshes, and mangrove swamps. Coastal habitats are valuable and varied, and they are often densely populated by humans. They are also the most threatened of all of the world's habitats.

Shallow coastal waters are well lit by the sun and richly supplied with nutrients. This combination of light and nutrients allows a rich growth of plants and animals that can be up to twenty times as productive as the open sea. But coastal environments also present difficulties. Their inhabitants must cope with alternating wet and dry conditions. An animal living at the top of the shore has to spend five or six hours out of water every tide. There is also a great variation in salinity, especially in estuaries where freshwater from rivers mixes with seawater.

Animals that must survive battering by waves inhabit rocky beaches. The main inhabitants include barnacles,

These forbidding cliffs, situated on Baccalien Island off Newfoundland in Canada, are home to so many seabirds that they are known as Bird Cliffs.

Coconut crabs grapple with each other in a battle for territory on a beach in the Togian Islands of central Sulawesi in Indonesia.

limpets, and sea anemones that cling to the rocks. Sandy beaches sometimes look as if they are empty. These beaches have fewer animals than rocky shores because less food is available. Animals must burrow in the sand to escape the powerful waves. Worms, crabs, shellfish, and other creatures lie under the sand and serve as food for birds at low tide and fish at high tide. Turtles and birds also make nesting places on sandy beaches.

Sandy beaches are popular for vacationers, and human activities pose the main threat to sandy beaches. Hotels and other buildings stretch for miles (km) along the best beaches in vacation spots. There is no room for wildlife when the beaches are covered with people. Sea turtles, for example, need undisturbed beaches for nesting. Tourist development of breeding beaches has serious consequences for the sea turtle population. Even in uninhabited places, beaches are damaged by oil and rubbish that is washed up by the tides. Serious environmental disasters have occurred when tankers wreck and their cargoes of oil spread over sandy and rocky shores.

Mangrove swamps cover about half of all tropical shores in the world. They occur as far north as Florida and China. Mangroves are small trees that grow in saltwater. Ocean currents carry their seeds, which take root in mud. To survive in this difficult environment, the trees grow buttress roots for support and special roots, called pneumatophores, that stick up through the thick mud and carry air to the roots that are buried deep in it. The leaves help get rid of excess salt that rain then washes away. The tangle of roots traps mud that, with the fallen leaves of the mangrove trees, provides the basis for life in the swamp. Some special mangrove animals include the mudskipper fish and several kinds of crabs that can live out of water for a long time and even climb trees.

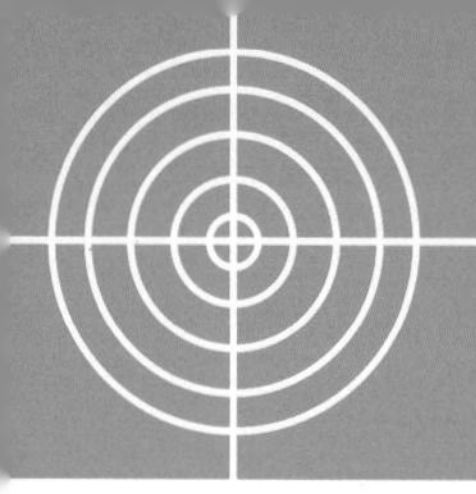

Mangrove swamps provide a living for millions of people. They protect coasts from storms, and the fertile, shallow water is a breeding ground for many kinds of fish, prawns, and oysters. These animals are very important for local people, who also use the mangroves and other trees. In southeastern Asia, the nypa swamp palm provides fruit, vinegar, alcohol, and fiber for thatching houses. Many mangrove swamps are disappearing because they are being turned into fish farms or rice paddies. Trees in some swamps are cut for firewood or turned into charcoal used for outdoor barbecues. People who live on tropical coasts must ensure that their mangrove swamps are preserved in order to protect their food supplies and prevent erosion. One tropical island had its mangrove trees cut down because mosquitoes bred in the swamp. The next hurricane, however, completely washed away the island. In Bangladesh, people and property were saved from a cyclone in 1991 because the coast had been planted with mangroves.

Salt marshes are the cool sea equivalent of tropical mangrove swamps. They occur on sheltered parts of coasts where mud accumulates, and they are very fertile. The upper part of a salt marsh is dry most of the time, but as it nears the sea, it floods at high tide. Only a few special plants, such as glasswort, sea lavender, and cord-grass, can survive the saltwater in these places. As the plants grow in the mud, they hold the mud together. Salt marshes provide winter homes for large flocks of birds that feed on the plants and small animals that live in the mud. Salt marshes are valuable because they act as

Chips of mangrove wood are ready to be shipped from Sanakan Sabah, Malaysia, to Japan as raw material for manufacturing rayon, paper, and other products. The removal of mangrove trees, however, can have devastating environmental consequences.

Waste washes up on the main beach at Puerto Plata in the Dominican Republic. Careless refuse disposal is a major threat to coastal and ocean habitats.

natural defenses against storms. They are destroyed when they are drained to make farmland. Only expensive seawalls can then keep the sea out.

An estuary is the place where a river widens and meets the sea. Mud brought down by the river gathers to form mud flats there. Sometimes, if the mud is not washed away, it spreads across the estuary, forcing the river to split into smaller streams that divide the mud banks into islands. This is called a delta. Salt marshes and mangrove swamps may form around the edges. Like the animals of the seashore, the inhabitants of an estuary learn to survive rapid changes in the saltiness of the water. The few species that inhabit estuaries are abundant, and they serve as food for a large number of animals. Estuaries are important nurseries for fish that spend their adult lives in open water. Two-thirds of the fish caught by the world's fishing fleets begin their lives in estuaries.

Estuaries make good harbors for large ships, so docks and factories are usually built around them. Reclaiming estuaries has been important for countries like the Netherlands that need more land for farming and towns. Unfortunately, when plans are made to develop estuaries, no notice is taken of their importance for the fishing industry. Damming the Nile River, for example, has resulted in the disappearance of fisheries around the delta. Life in estuaries can also be threatened by what happens many miles (km) up the rivers. Rivers often carry pollution as they move along, which has a serious effect on wildlife, and dams prevent mud from reaching the estuary.

CORAL REEFS

A close-up of coral encrusted with sponge. Coral reefs act as hosts to an enormous wealth of marine life.

Coral reefs occur in shallow water in tropical seas, where the water temperature is between 77°F and 84°F (25°C and 29°C). They are the marine equivalent of tropical rain forests because their wildlife is so rich and varied. And, like the rain forests, coral reefs are important for the health of the planet and for providing a living for the people who live near them.

Coral is made from the skeletons of millions of tiny animals, such as sea anemones. Coral animals have soft, round bodies with a mouth at one end surrounded by a ring of tentacles armed with stinging cells. Each coral animal forms a skeleton of calcium carbonate (lime) around its body. The skeletons of neighboring corals are often joined together to make branching colonies. When a coral animal dies, its skeleton remains, and new corals grow on top. Over hundreds of thousands of years, the skeletons accumulate to form a reef of solid limestone with a thin layer of living coral. Although corals are found in all seas, they form reefs only in tropical seas. As well as coral skeletons, a reef contains lime skeletons of seaweeds called coralline algae. They form a kind of cement over the reef that helps bind the coral together and protect it from storm damage.

The surface of the reef is home for a variety of seaweeds and every form of marine animal, including sponges, sea anemones, starfish, sea urchins, crabs,

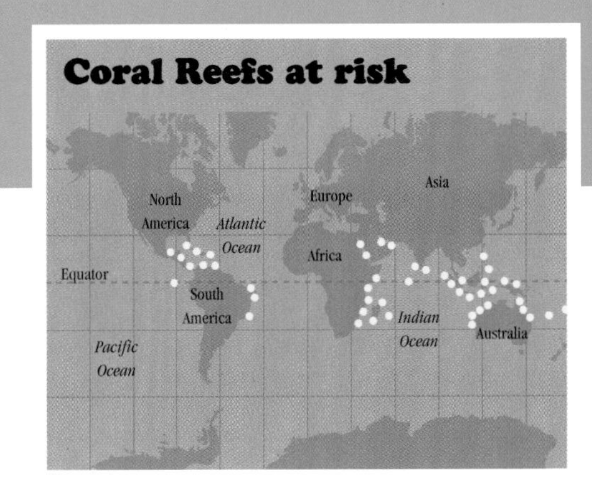

giant clams, and shoals of colorful fishes. The food source for all this life is plankton, tiny organisms that float in the sea. The corals and many other animals trap the plankton and they, in turn, feed larger animals. The coralline algae form a mat only a few inches (mm) thick, but it is the most productive part of the reef. Parrotfish, limpets, and sea urchins graze on the algae.

Fringing reefs, platform reefs, barrier reefs, and atolls are the main types of reefs. Fringing reefs and platform reefs grow in shallow water within a few miles (km) of the shore. Barrier reefs form farther offshore, at distances of up to 125 miles (200 km). The largest is the Great Barrier Reef of Australia, which is over 1,243 miles (2,000 km) long. Atolls are ring-shaped reefs that enclose a lagoon, often with an island at the center. Atolls may occur hundreds of miles (km) from the mainland. Some atolls grow on submerged volcanoes that have slowly disappeared below the sea while the reefs built up on them. Other atolls probably formed when the sea level was much lower during the Ice Age.

Coral reefs guard thousands of miles (km) of tropical shoreline. They protect the shore from storm erosion and allow the formation of sandy beaches and sheltered

A view of corals and other related fish life off the coast of Italy.

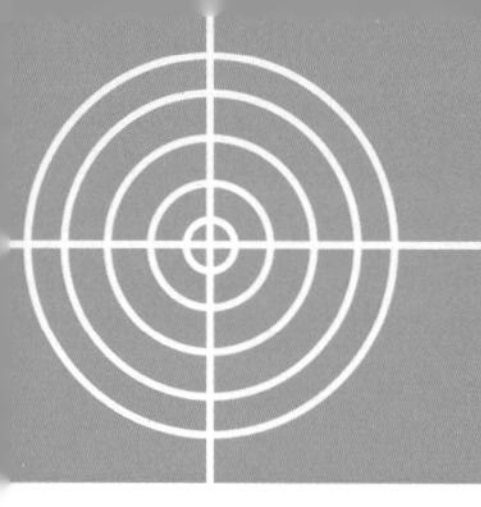

harbors. The reefs provide human inhabitants with rich harvests of fish, shellfish, and crustaceans, such as crabs and shrimps. Seafood is important on tropical islands that lack enough soil and water for agriculture. Coral is also used for building materials, and both coral and shells are used for ornaments and souvenirs.

Coral reefs can be damaged in many ways. Storms break up reefs and turn them into rubble. Changes in weather and ocean currents cause the water around the reefs to become too warm or too cold, and this kills the coral. These are natural events, and new growth of coral eventually replaces the lost reefs, but this can take many years. Sometimes human actions damage reefs so severely that the reefs eventually disappear. Other animals living on the reef vanish and, when the reef has broken down, the shore behind it is destroyed. This is currently happening in many tropical areas of the world.

Fringing reefs are easily harmed by human activities on shore. Coral reefs often grow where human populations are high, and seawater can be polluted from many sources. Pesticides and other pollutants are swept out to sea, where they contaminate reefs. When forests are cut down, soil washes down rivers and into the sea, where it smothers the corals. Reef fish are an important source of food for many people living on tropical shores. As the human population increases, there is a danger of overfishing. Some fishermen catch fish by exploding dynamite in the sea. This kills all the animals and

Coral blocks stand ready for use in the Maldive Islands in the Indian Ocean. Although coral is a traditional building material, overuse is causing severe damage to coral reefs.

Tourists can cause considerable damage by walking across coral reefs or interfering with marine life through scuba diving.

destroys the coral. Oil pollution is serious for reefs that lie uncovered at low tide. The reefs around the Arabian Gulf and the Red Sea are especially vulnerable because so much of the world's oil comes from the region. It is transported by tankers that often spill oil overboard.

Although coral reefs have been mined to produce building material for over two thousand years, the industry has now grown so large that it is a serious threat to reefs. Coral blocks are used to build hotels for tourists. The tourists have come to see the reefs and their animal life, but with this practice, the tourist industry is ruining itself. Coral is now banned as a building material in some places. Tourists help destroy the coral by walking on it at low tide and by collecting or buying pieces as souvenirs. Keeping tropical fish in aquariums is very popular. Most of the fishes come from coral reefs. Millions are bought every year, but three-quarters die, so more have to be caught. The process of catching the fish is also harmful to the reefs.

A big decline in coral reefs has occurred in recent years. About 10 percent have been damaged so severely they may never recover. One estimate is that two-thirds of all coral reefs will disappear in the next fifty to one hundred years. Coral reefs need the protection marine parks can provide. Corals and other species would survive there and spread back into damaged areas. Fishing and other activities must also be controlled if the reefs are to continue providing food and other resources for coastal people.

ISLANDS

Islands may form when a coral atoll provides a surface permanently above sea level that land plants and animals colonize over time.

Naturalists are interested in islands because the wildlife there is different from the nearest mainland. Fewer species of animals and plants live on islands than on the mainland, because it is difficult for mainland plants and animals to reach the islands. Island species are unique because they cannot be found elsewhere. Sometimes there are species, such as the tuatara from islands near New Zealand, that have become extinct in other places and now live only on islands. Other species, such as the takahe, have evolved on islands.

There are two main kinds of islands: continental and oceanic. Continental islands were once part of a large landmass but have become separated by the land sinking or the sea level rising. The animals and plants reached the islands before they were cut off. The longer the island has been separated, the fewer the species that will have reached it. The British Isles were once connected to the rest of Europe. Ireland has been an island longer than Britain, and it has fewer animals. The adder, mole, dormouse, wild cat, and roe deer live in Britain but not in Ireland. The mainland of Europe has many animals that do not live in Britain. The number of species also depends on the size of the island. In the Caribbean Sea, Cuba, the largest island, has ten times as many species of reptiles and amphibians as Montserrat, one of the smallest islands.

Oceanic islands have never had a connection with a continent. They were formed by volcanoes or coral reefs and are often hundreds of miles (km) from the nearest mainland. Iceland, Bermuda, Tristan da Cunha, the Galápagos Islands, and Hawaii are well-known oceanic islands. Most oceanic islands are in the Pacific Ocean, which contains nine hundred islands compared with only ninety in the Atlantic Ocean. The animals on an oceanic island must have come originally from land across the sea. Animals on the Galápagos Islands were brought by

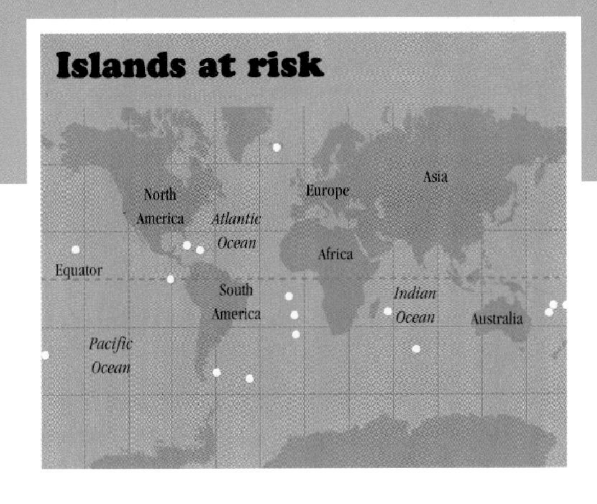
Islands at risk

an ocean current from the mainland of South America. The kinds of animals most likely to arrive on an island are those that can fly — such as birds, bats, and insects — or that can float. Large mammals are unlikely to reach islands, so there are usually no large predators. The result is that many of the inhabitants are very tame. Birds and insects may lose the use of their wings and become flightless. This is an advantage because they will not get blown out to sea while in flight. Some island animals grow larger than their mainland ancestors because there are no predators. The dodo of Mauritius was a turkey-sized relative of the pigeon.

In 1883, all animal life on the volcanic Indonesian island of Krakatoa was killed when the volcano erupted. Since then, scientists have monitored the arrival of new animal species from neighboring islands. Twenty-five years after the eruption, there were 13 bird species, 2 lizards, 192 insect species, and 2 snail species. After another 25 years, there were 1,100 species, including many more birds, 6 mammal species (mainly bats), 8 reptile species, and even an earthworm species.

After animals reach an island, they may evolve into new forms that are slightly different from their mainland ancestors. Charles Darwin's studies on the finches of the Galápagos Islands led him to his theory of evolution by natural selection. The finches were descended from a single species that had arrived from South America.

Carnivorous native snails are an ancient species that has survived on Poor Knights Islands off the coast of North Island in New Zealand.

Mining phosphates from guano. Overexploitation of island mineral resources can cause many environmental problems, including erosion and pollution.

Darwin noticed that every island had developed its own slightly different species of finches. Each type of finch had adapted to a slightly different environment.

Islands are living laboratories for studying evolution and ecology, but their wildlife suffers as soon as humans arrive. This happens for two main reasons. The animals are not accustomed to large predators, so they are easily caught and killed. Early visiting sailors and settlers killed animals, such as the dodo, for food. They also brought rats, cats, dogs, and pigs that hunted the native animals. Cats on Socorro Island, Mexico, completely destroyed the local dove population and nearly made the local mockingbird extinct. The habitats of the native animals were also destroyed when the vegetation was cleared for agriculture or eaten by introduced animals, such as goats and rabbits. These animals were sometimes released on islands as a food supply for any sailors that became shipwrecked. When rabbits were brought to Laysan Island, Hawaii, they destroyed the native vegetation in twenty years, and three of the five native land birds became extinct.

Most of the animals that have become extinct in the last few hundred years lived on islands. Of 90 birds that are now extinct, 80 lived on islands. Polynesians inhabited the Hawaiian Islands 1,600 years ago. When Europeans arrived 200 years ago, half the bird species were already extinct. Since then, more have become extinct, and most of the rest are endangered. The 18 species of flightless rails that once lived on islands, including the takahe of New Zealand, were easily killed by introduced predators. Only 7 species remain.

The size of islands and the small numbers of their inhabitants make island wildlife vulnerable, but this also

makes conservation easier. It is easier to get rid of pests from an island and restore island vegetation than to restore a large area of forest or grassland. In 1982, 10 goats lived on the island of Isabela, in the Galápagos Islands. There are now 100,000, and they are killing the forest home of the giant tortoises. In order to preserve this habitat and save the tortoises, the goats have to be destroyed. Other islands have been cleared of predatory cats and rats.

Round Island is a small island off the coast of Mauritius. It is the home of several species that are now extinct on Mauritius. About one hundred years ago, goats and rabbits were brought to Round Island. They reduced the forest to a single tree and the palm trees to ten. A rare gecko, a type of lizard, uses the palms to rest in during the day. All the goats and rabbits on Round Island have since been wiped out. The palms are growing again, and the gecko has a better chance of survival. Some of the native Round Island animals are also being kept in captivity until it is safe to return them to the island.

Humans are still a serious problem for islands, especially in the tropics. Native populations, for example, are increasing. After the disease malaria was controlled on Mauritius, the human population rose from 487,000 in 1950 to 1,100,000 in 1988. These people need to farm the land in order to survive, so there is less room for wildlife. Tropical islands are also very popular with tourists and provide much-needed money for the local populations. The Republic of the Maldives has as many tourists per year as permanent residents. New hotels, roads, golf courses, marinas, and the extra pollution caused by tourists all work to destroy the islands.

Charles Darwin's famous book, *The Origin of Species,* made the Galápagos Islands famous. These islands are home to a variety of rare and endangered species.

OCEANS

Oceans cover 70 percent of Earth's surface. They are, on average, 1.9 miles (3 km) deep, with some underwater valleys more than 2.5 miles (4 km) deep. The oceans contain 330 cubic miles (1,370 million cu. km) of saltwater, which is a gigantic store of nutrients for supporting a wealth of animal and plant life. Most life is concentrated in the top 98 feet (30 m) of the sea, where there is plenty of light, but some special animals live even at the greatest depths.

Although the oceans are linked together to make a single, huge mass of water, they are not the same. The waters are continually stirred by wind and tide into currents. The Gulf Stream is a major current that flows from the Caribbean Sea across the Atlantic Ocean to Europe and up the eastern coast of North America. It brings warm water northward. The Humboldt Current flows up the western coast of South America and takes cool water to the Galápagos Islands lying on the equator. Currents also carry surface water to the depths of the oceans, or from the depths to the surface. Currents are important for carrying nutrients, gases, and plants and animals to different parts of the ocean.

There are two main habitats in the ocean. The pelagic habitat is the water itself, and the benthic habitat is the seabed. The surface of the ocean, down to 328 feet (100 m) deep, is called the photic zone, which means enough sunlight exists for microscopic plants to grow. The most important plants are called diatoms. Tiny animals, especially crustaceans, eat the plants and are then eaten by larger animals, such as fish. The plants and the smallest animals that drift in the currents are called plankton. Larger animals that swim actively, such as fish and squid, are called nekton. The most productive parts of the ocean occur where there are enough nutrients,

A wandering albatross soars over the Southern Ocean.

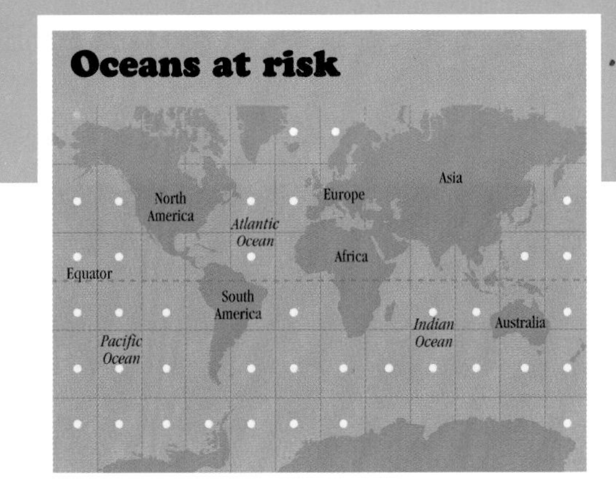

oxygen, carbon dioxide, and light for plant growth. Most ocean life is concentrated near the coast because nutritious deep water is brought to the surface by currents. Masses of planktonic plants grow on the nutrients and feed swarms of planktonic animals. Fish, squid, whales, seals, and seabirds gather to feed in these areas. Ninety percent of all fish are caught in coastal seas. The middle of the oceans are the equivalent of desert conditions because there is so little plankton for marine animals to eat.

Below the photic zone is the bathyal zone, where there is too little light for plants to grow. Life at the deep parts of the ocean depends mainly on oxygen carried from the surface by currents. Bathyal animals either eat the bodies of animals sinking from the pelagic zone, or they hunt each other. Many of the inhabitants are unusual-looking fish. They have weak bodies because there is very little food, but they are fierce hunters. When a meal appears, they must not miss it. Their huge mouths have long, sharp teeth that help them capture animals as large as themselves. Some bathyal animals, however, do not wait for prey to arrive. They swim to the surface to feed at night. The deepest water, below 6,560 feet (2,000 m) deep, is called the abyssal zone. It is cold and dark, and the pressure of the water is enormous. Only very special animals live there.

The benthic habitat, like the pelagic habitat, is richest in shallow water. Seaweeds grow around coasts, and nutrients are carried from the land by rivers flowing into the sea. Shellfish, worms, crustaceans, starfish, sea anemones, fish, and many other kinds of animals are common in these waters. Coral reefs also grow there.

Powerful ocean waves in Great Australian Bight near Koonalda, in South Australia.

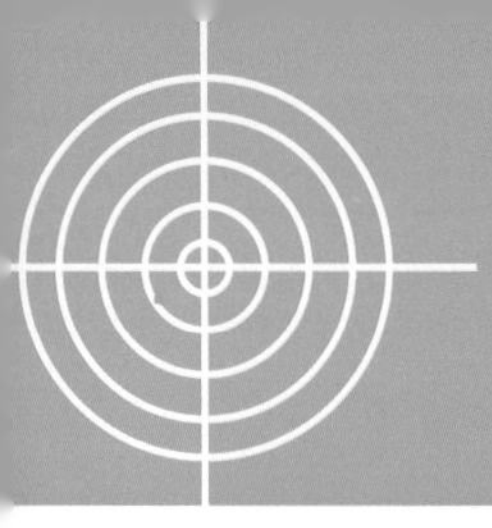

Oil escaping from the Braer oil tanker in the Shetland Islands. Fortunately, very heavy seas churning the water meant the environmental damage was less serious than originally anticipated. This is rare, and large-scale oil spills can cause years of environmental devastation.

Seaweeds do not live in deeper water, and the bottom-dwelling animals feed on the dead bodies of animals sinking from the surface waters. In the deepest places, the seabed is covered with fine mud, and animals feed by sifting nutrients from the particles.

People have been gathering food from the sea for thousands of years. As bigger boats were built, fishermen could go farther out to sea to find large shoals of fish. The best fishing grounds are in shallow water, such as the North Sea, the Grand Banks of Newfoundland, and the Gulf of Thailand. Large harvests of fish such as cod, herring, pilchards, and capelin have been very important for feeding growing human populations. At one time, it seemed as if there were an endless amount of fish to catch. But bigger fishing boats and bigger nets meant that too many fish were caught. Eventually, the size of the catches began to decrease, and so did the size of the fish that were caught. As the fish have run out, fishermen have had to search for new places to fish. Today, fleets of large fishing boats work in nearly every corner of the oceans. Many people around the world still fish from small boats to feed the local community, but their traditional livelihood is threatened by the large fishing boats. Their only protection lies in laws that prevent foreign fishermen from fishing in local waters.

The problem for the fishing industry is that no one owns the seas. Anyone can fish, and it is difficult to stop fishermen from catching too many fish. Some countries now legally claim the sea around their coast so they can control how many fish are caught. Yet, the increasing human population needs fish, so new fishing grounds and new kinds of fish have to be found. Around 110 million tons (100 million tonnes) of fish are caught every year, and many of the most valuable fish are disappearing.

Conservationists are worried that fishermen are killing many other animals, too. Some types of fishing nets are 37 feet (60 km) long, and they catch tens of thousands of seabirds, dolphins, turtles, seals, and also many species of fish other than the ones the fishermen want. Sometimes the fishermen lose these nets, but the nets continue to float in the sea for years and continue to catch and drown animals. These losses are especially devastating for animals that are already rare, such as the Amsterdam Island albatross and Kemps' ridley turtle

The history of whaling is similar to the history of fishing. When whalers killed all the whales in one part of the ocean, they found more to hunt at another location in the sea. When they killed off one species, they began to hunt another. Whales live in the open sea, so it is very difficult to control the whalers, and some species of whales have become very rare. They are now protected, but they are still being hunted.

Although the oceans are vast, they are becoming polluted even in the most distant points from land. Oil pollution is caused by leaks from ships' fuel tanks or by oil tankers being wrecked on the shore. Thousands of tons of oil spread over the surface of the sea, destroying life on beaches and killing seabirds, seals, and other animals that swim in the oil-covered water. In addition, chemicals are a problem when they are washed into the sea from rivers or when ships that carry them sink. Pollution is the most damaging in seas that are almost landlocked, such as the Mediterranean Sea and the Baltic Sea, where the chemicals cannot fan out.

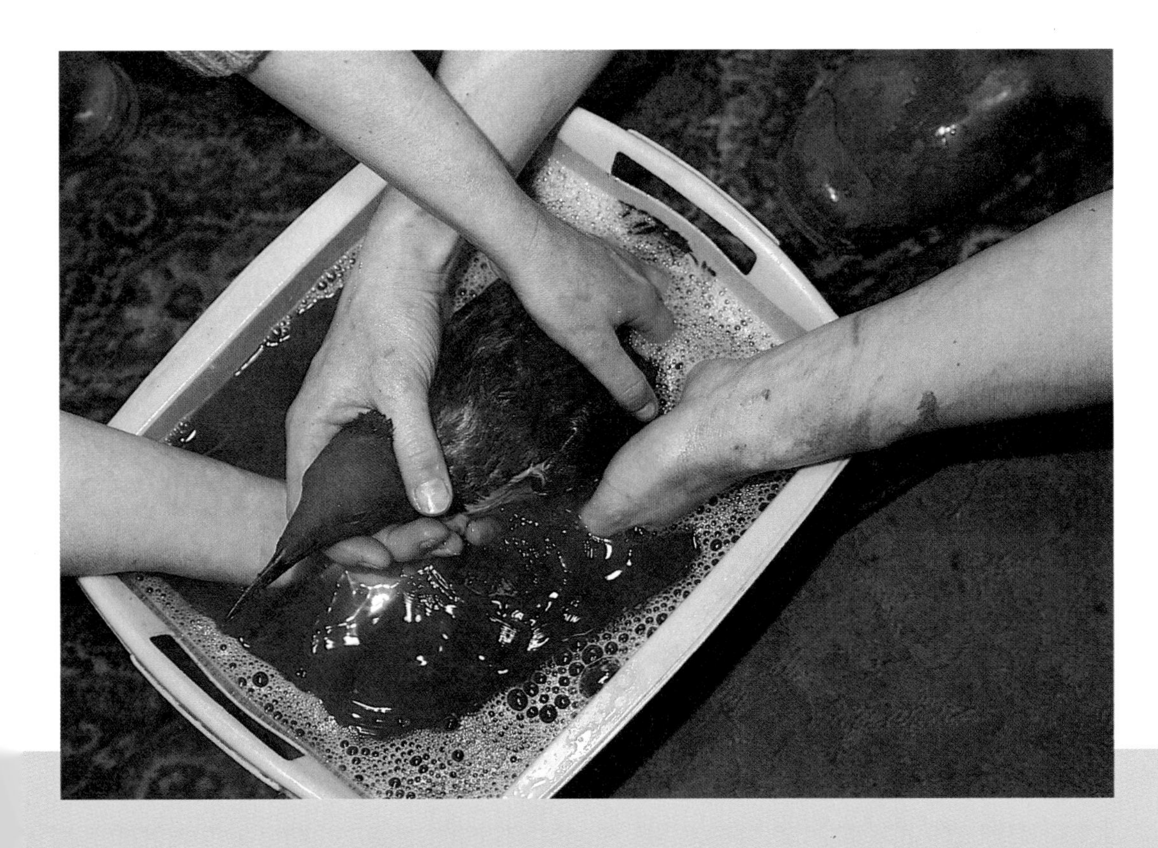

Cleaning up an oil-covered guillemot following an oil slick disaster off Cefn Sidan Sands in South Wales in January 1994.

GLOSSARY

acid rain — rain containing dissolved sulfur dioxide and nitrogen oxides, released into the atmosphere by the burning of coal or oil. This rain causes damage to vegetation and pollutes lakes and streams.

alpine — an animal, plant, or environment of the high mountains or alps.

alpine zone — the area on a mountain where only alpine plants can exist.

Antarctic — the area surrounding the South Pole. Antarctica is a continent surrounded by the Southern Ocean.

Arctic — the area surrounding the North Pole. No landmass exists at the North Pole. The Arctic ice-cap is frozen ocean.

avalanche — a mass of snow, ice, rock, and earth falling rapidly down the side of a mountain.

barren — infertile.

bathyal zone — the ocean habitat where the water is deeper than 328 feet (100 meters) and there is insufficient light for microscopic plants (phytoplankton) to grow.

benthic habitat — of or pertaining to life in the ocean depths.

biomes — natural communities of plants and animals covering large areas.

boreal forests — dense, coniferous forests in the Northern Hemisphere that extend from the Arctic tundra to the temperate forests.

carnivores — flesh-eating animals.

CFC — abbreviation for *chloroflurocarbon*, a chemical used in aerosols that some scientists believe causes damage to the ozone layer.

clear-cutting — the systematic removal of all trees from an area of forest or woodland.

climate — the pattern of weather conditions, such as rainfall, temperature, and wind, in a given area.

cloud forests — forests growing on mountainsides where the altitude and atmospheric conditions work to keep the forests almost permanently covered by clouds.

conservation — the act of preserving animals or plants from extinction.

continental plates — rigid plates that cover Earth's crust. They are also known as "tectonic plates." Volcanoes and earthquake zones are situated in regions where the continental plates move against each other.

crustaceans — invertebrate animals with segmented bodies and external skeletons or shells. Shrimps and crabs are crustaceans.

cuticle — the "skin" or film covering a plant.

deciduous trees — types of trees that shed their leaves during autumn.

deforestation — the act of cutting down trees and clearing forests.

desertification — the conditions that develop when humans overexploit land. The land then takes on the the characteristics of a desert.

deserts — dry environments that receive less than 9.8 inches (250 millimeters) of rainfall per year.

endangered — in danger of dying out.

epiphyte — a parasitic plant that grows on another plant for support.

equatorial — of or relating to the area of Earth around the equator.

erosion — the action of elements such as rain, wind, floods, rivers, and ice on Earth's surface, causing the gradual wearing away of rocks and earth.

evergreens — plants that keep their leaves and stay green throughout the year.

evolve — to change shape or develop gradually over a long period of time.

extinct — no longer in existence.

false deserts — deserts that have been created through desertification. Many desert species are unable to inhabit false deserts because of environmental damage.

floodplain — the lowlands adjoining a river that flood easily.

food chain — an arrangement of animals and plants in which each species is a food source for the next higher species in the series.

glaciers — masses of ice that move like slowly flowing rivers.

guano — the excrement of seabirds, which can build up to form vast deposits that can be mined to produce minerals and fertilizers.

habitat — the natural home of an animal or plant.

hibernation — passing the winter months in a state of rest in order to avoid extreme cold conditions and to conserve energy.

invertebrate — an animal that does not have a backbone.

irrigation — the process of carrying a supply of water to land (usually for agriculture) through artificial means, such as dams, pipes, channels, etc.

krill — shrimplike crustaceans that live in swarms and provide food for whales and other animals.

lagoon — a shallow lake, opening into a sea or river.

landlocked — an area or body of water that is almost or completely surrounded by land.

larva — in the life cycle of insects, amphibians, or fish, the stage that comes after the egg but before full development; e.g., a caterpillar is the larva of a butterfly or moth.

mammals — warm-blooded animals, including whales and dolphins, that feed their young on mother's milk.

mangroves — tropical trees that live in a waterlogged, salty environment.

marine — living or growing in the sea.

marsh — waterlogged ground.

migration — the seasonal movement of animals from one environment to another, often over long distances, and usually following certain routes.

monsoon — a wind accompanied by heavy rainfall in southern Asia that blows from the northeast in winter and from the southwest in summer.

montane — a mountainous environment.

native — originating in a particular place.

nomads — tribespeople who roam from place to place carrying their possessions and moving their herds of animals with them.

nutrient — any substance taken in by a living organism and used as a source of energy.

ozone hole — (*see* CFC) An apparent "hole" in the layer of Earth's atmosphere that protects our planet from the Sun's harmful radiation. This damage is often linked to the widespread use of CFCs.

pelagic habitat — of or pertaining to life in the open sea.

photic zone — the ocean habitat from the surface to 328 feet (100 m) deep, where sufficient light penetrates to allow microscopic plants (phytoplankton) to grow.

phytoplankton — the microscopic creatures in plankton that are plants, not animals.

plankton — drifting or floating forms of microscopic life (plants and animals) found in rivers, lakes, and oceans.

plantations — large estates of trees or shrubs that grow a single crop; e.g., tea, bamboo, sugarcane.

pneumatophores — special surface roots of trees, such as mangroves, that carry oxygen to the deep root system.

pollen — the fine, powdery substance produced by the male parts of flowers that fertilizes the female flower parts.

pollution — the gas, smoke, trash, and other harmful substances that damage our environment; e.g., oil spills and pesticides.

prairie — a grassy plain with no trees.

predators — animals that hunt and eat other animals for nourishment.

reptiles — cold-blooded animals with scaly skins; e.g., snakes, lizards, crocodiles, and tortoises.

reserves — areas of land set aside for the protection of wildlife.

salinity — the amount of salt in water or soil.

salinization — the process by which salt is left in soil by evaporating water. This makes the soil infertile.

salt marsh — an area of wet ground near the sea that floods at high tide.

savanna — an environment of grassy plains with clumps of trees.

species — a group of animals or plants whose members breed with each other, but do not breed with animals or plants outside the group.

steppe — a large, dry area of grassy plains with only a few trees.

succulents — plants with thick, fleshy leaves and stems that retain water.

swarm — a large group of insects that moves from place to place as a unit.

taiga — the Russian name for the belt of coniferous forest lying to the south of the Arctic tundra.

temperate — of or relating to the parts of Earth with a climate that falls between the heat of the tropics and the cold of the polar regions.

topography — a precise description of the surface characteristics of a place or area.

treeline — the altitude above which trees cannot grow. This varies in different locations according to local conditions, such as temperature and soil fertility. The treeline is also sometimes called the "timber line."

tropical — of or relating to the warm, humid area of Earth near the equator; the area of Earth that lies between the Tropic of Cancer and the Tropic of Capricorn.

tundra — a barren environment in which the subsoil is permanently frozen.

vertebrates — animals that have a backbone.

wader — a term for long-legged birds that wade, rather than swim, in marshes, lakes, or rivers.

MORE BOOKS TO READ

All Wild Creatures Welcome: The Story of a Wildlife Rehabilitation Center. Patricia Curtis (Lodestar)
The Californian Wildlife Region. V. Brown and G. Lawrence (Naturegraph)
Close to Extinction. John Burton (Watts)
Conservation Directory. (National Wildlife Federation)
Conservation from A to Z. I. Green (Oddo)
Discovering Birds of Prey. Mike Thomas and Eric Soothill (Watts)
Discovering Endangered Species (Nature Discovery Library). Nancy Field and Sally Machlas (Dog Eared Publications)
Ecology Basics. Lawrence Stevens (Prentice Hall)
Endangered Animals. John B. Wexo (Creative Education)
Endangered Forest Animals. Dave Taylor (Crabtree)
Endangered Grassland Animals. Dave Taylor (Crabtree)
Endangered Mountain Animals. Dave Taylor (Crabtree)
Endangered Species. Don Lynch (Grace Dangberg Foundation)
Endangered Species Means There's Still Time. (U.S. Government Printing Office, Washington, D.C.)
Endangered Wetland Animals. Dave Taylor (Crabtree)
Endangered Wildlife. M. Banks (Rourke)
Fifty Simple Things Kids Can Do to Save the Earth. Earthworks Group (Andrews and McMeel)
Heroes of Conservation. C. B. Squire (Fleet)
In Peril (4 volumes). Barbara J. Behm and Jean-Christophe Balouet (Gareth Stevens)
Lost Wide Worlds. Robert M. McClung (William Morrow)
Macmillan Children's Guide to Endangered Animals. Roger Few (Macmillan)
Meant to Be Wild. Jan DeBlieu (Fulcrum)
Mountain Gorillas in Danger. Rita Ritchie (Gareth Stevens)
National Wildlife Federation's Book of Endangered Species. Earthbooks, Inc. Staff (Earthbooks, Inc.)
Project Panda Watch. Miriam Schlein (Atheneum)
Save the Earth. Betty Miles (Knopf)
Saving Animals: The World Wildlife Book of Conservation. Bernard Stonehouse (Macmillan)
Why Are Animals Endangered? Isaac Asimov (Gareth Stevens)
Wildlife Alert. Gene S. Stuart (National Geographic)
Wildlife of Cactus and Canyon Country. Marjorie Dunmire (Pegasus)
Wildlife of the Northern Rocky Mountains. William Baker (Naturegraph)

VIDEOS

African Wildlife. (National Geographic)
The Amazing Marsupials. (National Geographic)
Animals Are Beautiful People. Jamie Uys (Pro Footage Library: America's Wildlife)
How to Save Planet Earth. (Pro Footage Library: America's Wildlife)
Predators of the Wild. (Time Warner Entertainment)
Wildlife of Alaska. (Pro Footage Library: America's Wildlife)

INDEX

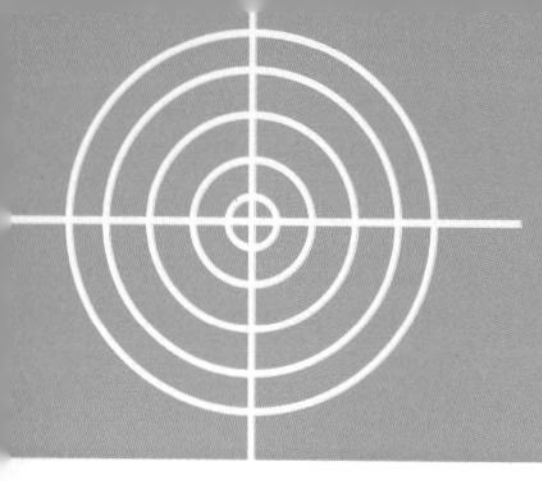

PICTURE CREDITS

Page 10, 11, 12, 15, 16 (upper), 20, 24, 25, 36, 42, 50, 52, Brian & Cherry Alexander
Page 13, 40, Marek Libersky/WWF
Page 14 Ph., Oberle/WWF
Page 16 (lower), 22, 28 (inset), 37, 53, Michael & Patricia Fogden
Page 17, Elizabeth Kemf/WWF
Page 18, 45, 49, Mark Rautkari/WWF
Page 19, Mark Hamblin/OSF
Page 21, André Maslennikov/WWF
Page 23, Wild-Type Productions/Bruce Coleman Ltd.
Page 26, Michael Gunter/WWF
Page 28, Mark Edwards/BIOS/WWF
Page 29, 43, 51, Gerald Cubitt/WWF
Page 30, 34, John Newby/WWF
Page 31, 39, Mark Boulton/ICCE
Page 32, Phil Chapman/ICCE
Page 33, Philip Steele/ICCE
Page 35, Tony Rath/WWF
Page 38, R. C. V. Jeffrey/WWF
Page 41, Jacqueline Sawyer/WWF
Page 44, Sylvia Yorath/WWF
Page 46, Bill Wood/Bruce Coleman Ltd.
Page 47, R. Rinaldi/Panda Photo/WWF
Page 48, Jack Stein Grove/Eye on the World/WWF
Page 54, Kim Westerskov/OSF
Page 55, Stephen J. Doyle/Bruce Coleman Ltd.
Page 56, Gryniewicz/© Ecoscene
Page 57, D. Halleux/Bios/WWF

WWF = World Wide Fund for Nature
OSF = Oxford Scientific Films
ICCE = International Centre for Conservation Education